Functional Groups: Characteristics and Interconversions

G. Denis Meakins

Emeritus Fellow, Keble College, Oxford

Series sponsor:  **ZENECA**

ZENECA is a major international company active in four main areas of business: Pharmaceuticals, Agrochemicals and Seeds, Specialty Chemicals, and Biological Products.

ZENECA's skill and innovative ideas in organic chemistry and bioscience create products and services which improve the world's health, nutrition, environment, and quality of life.

ZENECA is committed to the support of education in chemistry and chemical engineering.

OXFORD NEW YORK TOKYO
OXFORD UNIVERSITY PRESS
1996

Oxford University Press, Walton Street, Oxford OX2 6DP

Oxford   New York
Athens   Auckland   Bangkok   Bombay
Calcutta   Cape Town   Dar es Salaam   Delhi
Florence   Hong Kong   Istanbul   Karachi
Kuala Lumpur   Madras   Madrid   Melbourne
Mexico City   Nairobi   Paris   Singapore
Taipei   Tokyo   Toronto

and associated companies in
Berlin   Ibadan

Oxford is a trade mark of Oxford University Press

Published in the United States
by Oxford University Press Inc., New York

A catalogue record for this book is available from the British Library

Library of Congress Cataloging in Publication Data
(Data applied for)
ISBN   0 19 855867 8

Typeset by the author
Printed in Great Britain by
The Bath Press, Somerset

Series Editors Foreword

Functional group chemistry comprises a central part to modern organic chemistry courses. It provides the basic information necessary to understand elementary mechanisms and to design organic synthesis strategies to simple targets.

Oxford Chemistry Primers have been designed to provide concise introductions relevant to all students of chemistry and contain only the essential material that would normally be covered in an 8–10 lecture course. In this primer Denis Meakins presents a systematic and student-friendly introduction to functional group chemistry that will give students a firm grounding in basic reactions and reactivity at the beginning of their chemistry courses. This primer will be of interest to apprentice and master chemist alike.

Stephen G. Davies
The Dyson Perrins Laboratory, University of Oxford

May 1996

Preface

The characteristic properties of functional groups and the methods for interconverting them are the foundations of organic chemistry; a sound grasp of these topics is essential for the aspiring chemist's journey to the higher levels of the subject.

It is probably in the field of synthesis that the importance of functional group interconversion shows up most clearly. The approaches to new compounds (required, perhaps, for their physiological activity) involve many stages in which a precursor's functional group must be transformed into the one required for the next stage. To effect such changes efficiently is the hallmark of a successful synthesis.

All the information about functional groups is already in many excellent books and the need for a Primer may, therefore, be questioned. The modern text-books are long (ca. 1000 pages); they contain the functional group material and a great deal more! A student wanting information on say, ketones is faced with twenty or more page references, and to wade through these without guidance is a daunting or even an overwhelming task. The object here is to anticipate this difficulty by presenting the chemistry of the groups in a *concise* and *systematic* form.

To deal with functional groups in so short a compass as the present one necessitates difficult decisions about what must be included and what may be omitted, even though reluctantly. The present selection has been formed by experience in giving the first year lecture course; I am grateful to the undergraduates for many helpful comments and particularly to the Keble chemists with whom I enjoyed a rewarding relationship as their Tutor for many years.

A colleague of long standing, Dr A. S. Bailey (Emeritus Fellow of St Peter's College), kindly read the first draft, spotted mistakes and made many helpful suggestions. I am grateful for his encouragement and generous support.

Oxford
May 1996

G. D. M.

Contents

The Schemes are on, or start on, the following pages

1 The approach, and background topics

1.1 Introduction

Although most of the material is for study in the first year at university, some of it belongs more appropriately to later years. Topics regarded as beyond the first year are clearly identified in the text by the superscript[#]. These will be covered in the second year, and it is hoped that the book as a whole will be useful for quick reference and revision at all stages. Some pages are shorter than the standard length; these end, irrespective of their length, at natural breaks between topics.

The aim is to cover the characteristic properties of functional groups and the methods for interconverting them in a *concise systematic* form. Only those properties and reactions regarded as general and important are included. A *concise* treatment requires a very economical presentation. The form adopted consists of schemes and diagrams, very short sentences and (in many places) a 'note form' of key phrases. Grammatical correctness and style have thus been subordinated to brevity and clarity. The central feature of the attempted *systematic* treatment is the simple reference system described a little later.

Aliphatic compounds (those not containing rings of atoms, open-chain compounds) are the main subject but cyclic compounds (those containing a ring) are often used to give a clearer illustration of a point. A few aromatic compounds are included to illustrate similar behaviour of aliphatic and aromatic compounds arising from a common structural feature.

The rest of this chapter is concerned with the terms, representations and abbreviations which are to be used, and with some fundamental topics (e.g. electronic effects) involved in studying the functional groups. Although it is not necessary for us to delve too deeply into organic theory a knowledge of the main features is required. Most students will already have some familiarity with the material; helpful accounts, pitched deliberately at an elementary level, are given in another Primer (M. G. Hornby and J. M. Peach, 'Foundations of Organic Chemistry', abbreviated here to 'Foundations'). In most departments the study of functional groups is preceded or accompanied by a course on theory and mechanism. However, this supplementation cannot be assumed, and to embark on the groups without some knowledge of certain fundamentals would be difficult and unsatisfactory. To fill such a possible gap, brief accounts of the topics germane to group chemistry are given in the following sections. For simplification the valence bond treatment is generally used, but simple molecular orbital theory is introduced at certain points.

1.2   The functional groups, and order of discussion

The groups to be discussed are shown in scheme **1.2Ge**.

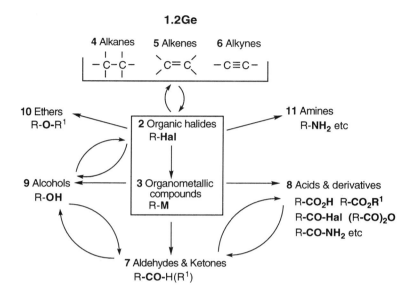

1.2Ge

The organic halides (**2**) involving the chemistry of $C^{\delta+}$ are discussed first. Next come the organometallic compounds (**3**) prepared from the halides; these exemplify the contrasting situation, the chemistry of $C^{\delta-}$. Thus, the behaviour of two important states of C is encountered at an early stage. From these two central groups the chemical cycle, which embraces the others, is developed in chapters following the numerical order of the scheme. [There is no group (**1**); this number refers to the present chapter.] It is thought that a mechanistic understanding emerges more naturally from the order adopted here than from the traditional sequence starting with hydrocarbons.

Transformations, arrows in 1.2Ge, signify that while some conversions are largely restricted to one direction some pairs, e.g. acids and aldehydes, can be usefully interconverted. Systematic names (see Chapter 4) are given, in plain (normal) type, for almost all the individual compounds. However, with some compounds the traditional names, given in *italic type*, are so well established that there is little hope of change. For groups of compounds the name most clearly reflecting their nature is used even if there is a systematic alternative. For example, the main compounds in the first group to be considered are called alkyl halides rather than halogenoalkanes.

You may like to skim through the rest of this chapter to get the gist of the material. Then start in earnest on Chapter 2. When you come to a point (such as a term, an abbreviation, an effect etc.) whose meaning is not clear turn back to the appropriate section here and study it in more detail.

1.3 Reference system

Only chapter 1 is divided into many (13) sections. Of the other chapters some are divided into sections but most are not divided. The chemistry is regarded as General (abbreviated **Ge**), Preparations (**Pr**) and Reactions (**Re**). Where necessary **Ge**, **Pr** and **Re** are further divided, as shown by the examples in the margin. **Ge** includes topics which apply generally, e.g. the inductive effect, and also the main tendencies of the individual functional groups (which are summarised at the start of each chapter or section). This simple system allows reference forward to material coming later and backwards to that already covered.

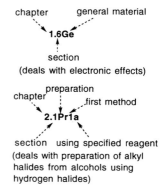

1.4 Representations in schemes

R = (means) unbranched **alkyl** group C_nH_{2n+1} e.g.

methyl (CH_3 or Me)   ethyl (CH_3CH_2 or Et)   propyl ($CH_3CH_2CH_2$ or Pr)   butyl ($CH_3CH_2CH_2CH_2$ or Bu)

The properties of a functional group are often influenced by the nature of the α carbon atom, the atom to which the group is attached. When it is necessary to show this influence the representation of the R group is expanded as follows:

β α
C— C— functional group

RCH_2Br	R_2CHBr	R_3CBr
primary alkyl group	secondary alkyl group	tertiary alkyl group
abbreviated to **pm**	abbreviated to **se**	abbreviated to **te**

Ar = **aryl** (aromatic) e.g. **Ph** = phenyl, **PhCH$_2$** = benzyl, **Ts** = abbreviation derived from traditional name *toluenesulphonyl*

systematic name 4–methylbenzenesulphonyl

[Nomenclature (naming compounds) comes later in **4.1Ge**]

Ph–   Ph–CH$_2$–   Ts–

R— C = **acyl** group, often represented **R–CO–**   Ac – = Me – C

Most common Me–CO– often represented Ac– from *acetyl,* (ethanoyl)

Hal = **halogen** atom; in many Re certain halogens are specified as more suitable than others

M = **metal** atom; most common for organic work Li, Mg, Cd, Al

'**Free bonds**' (nothing at one terminus) go to **C** or **H** unless specified otherwise e.g.

C=O includes   C=O   C=O   C=O   C=O   but not   C=O

aldehydes & ketones   alkyl groups: identical, different   acids

1.5   Reactions

Many reactions are of the form:

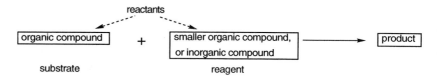

Terms used:

Δ = heat; usually 60–80° (degrees° are °C throughout)

sq = small quantity of a reagent; > catalytic amount, about 1–5% by weight of substrate

ts = transition state (see 'Foundations'); used only when no ambiguity possible

(For the present don't try to distinguish between the next two; regard them as the *slow* step of a multi-step sequence)

rds = rate-determining step; the first step of a multi-step sequence which is slower than the subsequent steps

rls = rate-limiting step; the later step of a multi-step sequence which is slower than the others

Kinetic and Thermodynamic control: Consider a reaction which may give more than one product. If the main product is the one formed *faster* (of two products) the reaction is said to proceed under kinetic control. If the main product is the *more stable* (of two products), thermodynamic control. Generally there is no problem, the product formed faster is the more stable. Only *reversible* reactions can be subject to thermodynamic control.

In the following example the chemistry is difficult and need not be considered here; the crucial point is that a reaction gives different products at different temperatures. Product **P** (lower activation energy, higher rate constant) is formed irreversibly at 20°. At 80° sufficient thermal energy is available for equilibria to be set up between **P** and reactants, and between reactants and **Q**. **Q** gradually accumulates until the equilibrium position between **P** and **Q** governed by ΔG_0 is reached.

k_1, k_2 are rate constants ; $\Delta G_1^{\ddagger}, \Delta G_2^{\ddagger}$ are activation energies; ΔG_0 is the energy difference, **P–Q**. **P** is formed under kinetic control, and **Q** under thermodynamic control

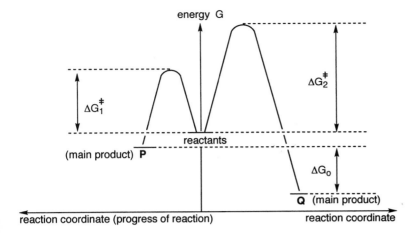

1.6   Electronic effects, resonance

Resonance in relation to benzene is discussed in 'Foundations'. Representations of electronic distribution in terms of resonance and of the mesomeric effect (mesomerism) are alternatives; they arrive at the same result, as illustrated in the example later in this section.

Inductive effect (**I**): Sign and order, see **1.6Ge**. This arises from the difference in electronegativity of an atom or a group relative to H (hydrogen). It affects σ-electrons in a covalent bond, and is represented by an arrow on the bond. The effect results in a small separation of charges (polarisation). It falls off rapidly with distance (see βC atom).

e.g. $C \rightarrow Hal$ leads to $\overset{\delta+}{C} - \overset{\delta-}{Hal}$

Hal has a $-I$ effect

$\overset{\delta\delta+}{C} - \overset{\delta+}{C} - Hal$
$\;\beta\quad\;\alpha$

1.6Ge

−I   most neutral and positive groups $(F > Cl > Br > I)$

$C \rightarrow \overset{+}{N}H_3 > C \rightarrow NO_2 > C \rightarrow CN > C \rightarrow CO_2H >$ $C \rightarrow Hal$ $> C \rightarrow OR > C \rightarrow C \overset{O}{\underset{R}{\diagup\!\!\backslash}} >$

$C \rightarrow SR > C \rightarrow C \equiv CR > C \rightarrow Ar > C \rightarrow C\underset{\underset{C-}{\diagdown\!\!\diagup}}{\diagdown}$

+I   negative groups, and alkyl groups $(R \rightarrow C$ relative to $H - C)$

$\overset{-}{S}, \overset{-}{O} \rightarrow C > Me_3C \rightarrow C > Me_2CH \rightarrow C > MeCH_2 \rightarrow C > Me \rightarrow C > D \rightarrow C$

+M $C \overset{\curvearrowleft}{NR_2} > C \overset{\curvearrowleft}{OR} > C \overset{\curvearrowleft}{SR} > C \overset{\curvearrowleft}{Hal}$ $(F > Cl > Br > I)$

−M $C = \overset{+}{N}R_2 > C = S > C = O > C = NR > C = CR_2$

Keep in mind that $F > Cl > Br > I$, and that $OR > SR$ in *both* $-I$ and $+M$

Dimethylformamide
N,N–dimethylmethanamide

Resonance

Mesomeric effect

Composite representation

−**M**, becomes negative

+**M**, becomes positive

follow arrows
through

Resonance: This is the concept invoked when the properties of a compound cannot be represented satisfactorily by one conventional formula. The structure is regarded as being somewhere between two or more formulae. For example dimethylformamide (see margin) does not behave as either a normal aldehyde or a normal amine. Formula P is therefore not a satisfactory representation.The structure is somewhere between P and Q, *hypothetical* extreme forms called *canonicals*. Canonicals differ in only electron distribution; the atomic centres are in the same places.The compound is termed a resonance hybrid, resonance being depicted by the double-headed arrow shown. The energy of the compound (its enthalpy) is lower than the energy calculated for any of the canonicals. Resonance is *not* a mechanical oscillation between real forms.

Mesomeric (also termed Resonance or Conjugative) *effect* (**M**): Sign and order, see **1.6Ge**.This involves interactions of π electrons, or of π electrons and unshared electron pairs (also termed nonbonded or lone pairs).It results in changes in covalency of two or more centres, and is represented by curly arrows. This effect does not fall off with distance (see aromatic example). Operation of the **M** effect with e.g. dimethylformamide suggests a structure between P and Q, the same outcome as reached on the resonance concept.

1.7 Acids and bases

There are two types of acids. The first type (Bronsted acids), e.g. $MeCO_2H$, give an H^+ to the substrate. The second type (Lewis acids), e.g. BF_3, coordinate with an unshared electron pair in the substrate. Similarly, Bronsted bases, e.g. NMe_3, accept an H^+ from the substrate, and Lewis bases, e.g. Et_2O, provide an unshared pair for coordination to the substrate. We shall be concerned mainly with the Bronsted type, represented **H–A**(acids) and **B**(bases).

The strengths of H–A are expressed on the pK_a scale

e.g. $MeCO_2H + H_2O(excess) \rightleftharpoons MeCO_2^- + H_3O^+$

ethanoic (*acetic*) acid

$K_a = [MeCO_2^-][H_3O^+] / [MeCO_2H] = 1.7 \times 10^{-5}$ mol dm^{-3} at 298K

$pK_a = -\log_{10}K_a = 4.8$

([H_2O] stays almost constant at 55.6 mol dm^{-3} and is incorporated into K_a.)

In order to have a unified scale for H–A and B strength, the acid strength of BH^+ rather than the base strength of B is listed. If required, the latter value is easily calculated from the former, as shown in the following example:

For $MeNH_2$ acting as a base require pK_b. This is $-\log_{10}K_b$ in the equilibrium

$MeNH_2 + H_2O \rightleftharpoons MeNH_3^+ + OH^-$

The literature lists the acid strength of the $MeNH_3^+$, the *conjugate* acid of $MeNH_2$ (conjugate means related) in the equilibrium

$MeNH_3^+ + H_2O \rightleftharpoons MeNH_2 + H_3O^+$ $pK_a (MeNH_3^+) = 10.6$

In any aqueous solution $[H_3O^+][OH^-] = 10^{-14}$mol^2 dm^{-6} at 289K, and by writing out the expressions in full it can be shown (see 'Foundations') that

pK_b(of B) + pK_a(of BH$^+$) = 14

Thus $pK_b(MeNH_2) = 14 - 10.6 = 3.4$

As the acid strength of H–A *increases* K_a increases and pK_a *decreases*. As the base strength of B *increases* K_b increases and pK_a (of BH$^+$) *increases*.

A cautionary note: two values may be listed for what is, ostensibly, the same compound, e.g. 4.6 and ~27 (~ means 'about') for PhNH$_2$. However pK_a 4.6 refers to   PhNH$_3^+$ + H$_2$O $\rightleftharpoons$ PhNH$_2$ + H$_3$O$^+$ whereas pK_a ~27 refers to PhNH$_2$ + H$_2$O $\rightleftharpoons$ PhNH$^-$ + H$_3$O$^+$

The second process is a convenient fiction. PhNH$_2$ is a base not an acid; formation of PhNH$^-$ from PhNH$_2$ requires a very strong base.

Aid to thinking: pK_a of H–A = pH of a solution in which [H–A] = [A$^-$]. Thus for MeCO$_2$H   [MeCO$_2$H] = [MeCO$_2^-$] at pH 4.8

1.7Ge shows a selection of pK_a values which will be useful later.

1.8   Electrophiles, nucleophiles, radicals

An electrophile (electron seeking), represented by **E$^+$**, has a centre (usually of low electron density) which forms a bond to the substrate using the substrate's electrons. A nucleophile (nucleus seeking), represented by Nu$^-$, has a pair of electrons (unshared or in a π bond) which form a bond to a centre of low electron density in the substrate; for most organic reactions the centre is a C atom with some degree of positive charge. A radical, represented by **U·**, has an unpaired electron and is usually very reactive. Some reagents, e.g. HBr, can act either as (Bronsted) acids or electrophiles ; some, e.g. NH$_3$, as bases or nucleophiles.

1.9   Carbocations, carbanions, carbon radicals

Carbocations are planar; stabilised by electron donating groups, destabilised by electron withdrawing groups. *Carbanions* may be regarded as planar. ($^\#$Actually mixtures of rapidly inverting tetrahedral forms, as in NH$_3$.) Carbanions are stabilised by electron withdrawal and destabilised by electron donation. *Carbon radicals* may be regarded as planar. ($^\#$Some deviate slightly from planarity. The situation is similiar to that of carbanions but with 'shallower' tetrahedra.) Their relative stabilities, which cannot be explained in terms of inductive effects of attached groups, correlate with the variation of C–H homolytic bond energy as H atoms are replaced by R groups (see **4.2Re1a**).

1.9Ge

Stability order	te (tertiary)		se (secondary)		pm (primary)
Carbocations	R$_3$C$^+$	>	R$_2$CH$^+$	>	RCH$_2^+$
Carbanions	R$_3$C$^-$	<	R$_2$CH$^-$	<	RCH$_2^-$
Carbon radicals	R$_3$C·	>	R$_2$CH·	>	RCH$_2$·

1.7Ge

acid strength increasing

Acids	pK_a
HCl	−7
CCl$_3$CO$_2$H	0.9
ClCH$_2$CO$_2$H	2.8
PhCO$_2$H	4.2
MeCO$_2$H	4.8
O$_2$N–⟨⟩–CO$_2$H	7.2
HCN	9.2
PhOH	10

Neutral	
H$_2$O	15.7 (not 14)
EtOH	18

Bases		pK_a of BH$^+$
O$_2$N–⟨⟩–NH$_2$	19	1.0
PhNH$_2$	~27	4.6
pyridine		5.3
NH$_3$	~36	9.2
MeNH$_2$	~37	10.4
piperidine		11.2
OH$^-$		15.7

base strength increasing

E$^+$ may be +ve or neutral e.g. H$^+$, NO$_2^+$, Hal$_2$

Nu$^-$ may be −ve or neutral e.g. MeO$^-$, NH$_3$

U· may be organic or inorganic e.g. Me·, Cl·

carbocation

carbanion

radical

1.10 Substitution, elimination, configuration

Many substitution reactions are of type:

R—L + nucleophile ⟶ R—nucleophile + L⁻

L⁻ is the *leaving group* and may be negative (e.g. Br⁻) or neutral (e.g. Me_3N).

The first part of the following scheme depicts substitution and elimination reactions from a system in which L is attached to a primary alkyl group; in the second part (next page) L is attached to a tertiary alkyl group. The chemistry is discussed in the text which comes after the scheme.

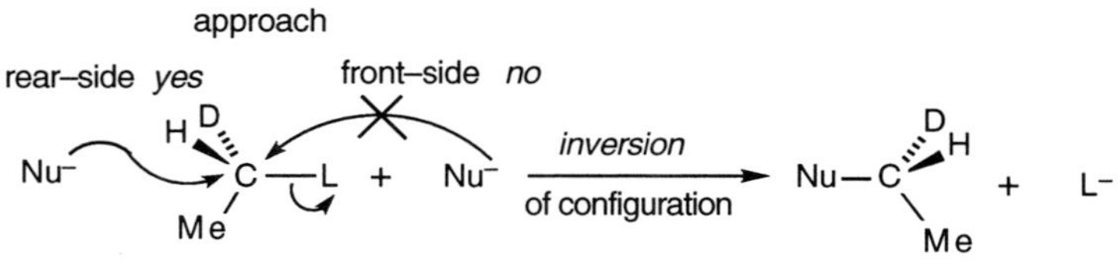

[The statements *yes* and *no* used here and elsewhere refer to the main outcome. Most reactions give mixtures, so *yes* does not imply entirely one product or process]

S$_N$2 $\underline{S}$ubstition by a $\underline{n}$ucleophile in a $\underline{b}$imolecular reaction (Avoid defining as second-order reaction)

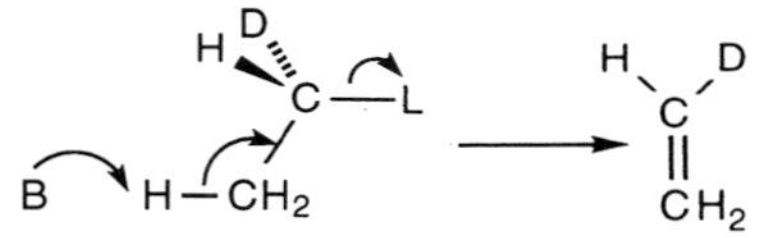

E2 $\underline{E}$limination in a $\underline{b}$imolecular reaction

#The complete definitions of S$_N$2 and E2 are that the slow step (rds or rls) is bimolecular. Here there is only one step in both which, therefore, must be the rds.

Ar— = MeO—⟨benzene⟩—

Nu⁻ difficult

marked hindrance

Ph, Ar, C—L

rds (slow)
(←———)

Me

Ph, + Ar
C
|
Me

+ L⁻

50% from above   50% from below

Nu⁻

Sₙ1 Substition by a <u>n</u>ucleophile, rds (slow step) is <u>uni</u>molecular

fast

Ph, Ar, C— Nu
Me
retention

Ar, Ph
Nu— C
Me
inversion of configuration

equal amounts, process termed *racemization*

marked hindrance

Ph, Ar, C—L
difficult
H— CH₂
B

side view of the carbocation

Ph   Ar
+, C
much less hindrance
|
B   H—CH₂

fast

Ph, C, Ar
‖
CH₂

E1 <u>E</u>limination in which rds (slow step) is <u>uni</u>molecular

To follow the stereochemistry of the reactions chiral substrates must be considered. (Chirality is discussed in 'Foundations'.) Thus to illustrate primary substrates Me–CHD–L (rather than Me–CH₂–L) is used because, by virtue of the D atom, this molecule has an asymmetric C. In the Sₙ2 (*substition*) reaction the C has undergone an 'umbrella' motion. The C–Nu orbital is not the original C–L orbital. The reaction involves *inversion of configuration*. In front-side approach Nu⁻ and L would 'get in each others way'. Rear-side approach involves much less steric hindrance. (#It is also strongly favoured by a stereoelectronic factor.) For comparison with the following tertiary case keep in mind that a carbocation Me–CHD⁺ would be very unstable. In the E2 (*elimination*) reaction removal of the H is concerted with (at the same time as) departure of L⁻.

The chiral tertiary compound exemplifies Sₙ1 and E1 reactions. Sₙ2 and E2 are impeded by severe steric hindrance to approach of nucleophiles or bases from either side. A different pathway involving ionisation is followed. The carbocation formed is much more stable than that from a primary compound but still very reactive. (#The first step is reversible but the backwards reaction is so much slower than the second step that it is usually not significant.) The carbocation common to Sₙ1 and E1 is planar; the top and bottom faces are therefore equivalent. Sₙ1 leads to *racemisation*.In E1 removal of the H occurs after the departure of L⁻.

In general, 'other things being equal', the most favoured reactions are S_N2 for primary and E1 for tertiary. However many factors come into play and these tendencies must not be regarded as rules. For example by appropriate choice of reactants and conditions tertiary compounds may be induced to undergo S_N1 or even S_N2 reactions (**2.1Re3b,c**).

1.11 Leaving group, acid strength: Nucleophilicity, base strength

This section is based on S_N2 reactions in a *protic* solvent (see Section 1.13).

$$R—L \ + \ Nu^- \longrightarrow R—Nu \ + \ L^-$$

The relative rates of series of the reactions give the tendency of an entity to act as L^- (good.......poor) and as a nucleophile, Nu^- (strong.......weak, termed its nucleophilicity). Several investigations, under different conditions, have given somewhat different sets of results. Although the *figures* for the relative rates in **1.11Ge** are not important, the *orders* of the L^- and nucleophile species are very useful.

The following tempting arguments suggest themselves. As the acid strength of H–L increases the stability of L^- increases, so L^- should become a better leaving group. Thus leaving group tendency should parallel acid strength. Similarly, as the basicity of a species (tendency to accept H^+) increases, its nucleophilicity (tendency to attack $C^{\delta+}$) should increase. Thus basicity should parallel nucleophilicity. *There are no such general correlations*. Several factors militate against the existence of such correlations. The main one is that acid and base strengths refer to *equilibria* (difference in free energy between ground states) whereas leaving group tendency and nucleophilicity represent *rates* (difference in free energy between a ground state and a transition state).[[#]Full discussion, later course on 'Linear Free Energy Relationships'.]

There are some useful correlations for species of similar structures (margin) but even these are not immutable. Fortunately, choosing an L^- or a Nu^- for a particular purpose is facilitated by the wealth of experimental results available in extended versions of **1.11Ge**.

Nucleophilicity and basicity in protic solvents (**1.11Ge**)

$EtO^- > PhO^- > MeCO_2^-$

$I^- > Br^- > Cl^- > F^-$

but as Nu^- in aprotic solvents

$F^- > Cl^- > Br^- > I^-$

see Section 1.13 for this reversal in nucleophilicity

1.11Ge

The relative rate figures are approximate; they vary in different S_N2 reactions

	L^-	Relative rate	Conjugate acid,	and its pK_a		Nu^-	Relative rate	Conjugate acid,	and its pK_a
good	N_2	high				PhS^-	5×10^5	PhSH	7
good	TsO^-	6	TsOH	−6.5	v strong	CN^-	5000	HCN	9.3
good	I^-	3	HI	−10	v strong	I^-	5000	HI	−10
medium	Br^-	1	HBr	−9	strong	N_3^-	1000	HN_3	4.7
medium	H_2O	1	H_3O^+	−1.7	strong	$PhNH_2$	1000	$PhNH_3^+$	9.4
medium	Me_2S	0.5	Me_2SH^+	−5.2	strong	EtO^-	1000	EtOH	18
poor	Cl^-	0.02	HCl	−7	strong	OH^-	1000	H_2O	15.7
poor	F^-	0.001	HF	−3.2	medium	Br^-	500	HBr	−9
rarely act as L^-	OH^-	~0	H_2O	15.7	medium	PhO^-	400	PhOH	10
rarely act as L^-	NH_2^-	~0	NH_3	~36	weak	pyridine	50	pyridinium	5.3
					weak	$MeCO_2^-$	50	$MeCO_2H$	4.8
					weak	Cl^-	50	HCl	−7
					not useful	F^-	<1	HF	−3.2
					not useful	TsO^-	<1	TsOH	−6.5
					not useful	H_2O	<1	H_3O^+	−1.7

1.12 Steric effects on basicity and nucleophilicity

Preoccupation with electronic effects should not be allowed to result in neglect of steric effects; these are central to all branches of organic chemistry. A general consideration and some important manifestions are shown in **1.12Ge**.

In general, attack by a nucleophile is sterically more demanding than attack by a base (top part of scheme). The contrast between primary and tertiary alkoxides (middle of scheme) has significance in the preparative work of later chapters. More subtle features are exemplified in the material lower down the scheme.

1.12Ge

Substrate, represented:

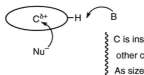

Nu⁻ → C$^{\delta+}$—H ↙ B

} C is inside molecule, approach of reagents may be impeded by
other centres. H is on periphery, less steric hindrance to approach
As size of reagent increases the importance of this difference
increases

pK_a values (conjugate acids)

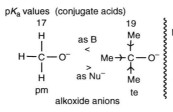

```
        17              19
        H               Me
        |      as B      |
 H — C — O⁻    <    Me → C — O⁻
        |      >         |
        H    as Nu⁻      Me
       pm              te
```

alkoxide anions

} Me groups (−I effect) push electrons towards O which is
already −ve. Anion destabilised, hence stronger base.
Me groups increase size of reagent (which becomes
fatter), hence lower nucleophilicity

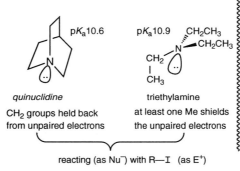

pK_a10.6

quinuclidine

CH₂ groups held back
from unpaired electrons

pK_a10.9

```
        CH₂CH₃
   N ◀—CH₂CH₃
  CH₂
   |
   CH₃
```

triethylamine

at least one Me shields
the unpaired electrons

reacting (as Nu⁻) with R—I (as E⁺)

} Basicities about the same, little
difference in hindrance with H⁺
(small). But relative rates,
quinuclidine: triethylamine, in
reactions with E⁺ are:
MeI as E⁺, 50 : 1, marked difference
Me₂CHI as E⁺, 500 : 1, huge difference.
Approach more hindered; therefore
enhancement of difference in
accessibility of unpaired electrons

1.13 Solvents

Unfortunately many overlapping terms are used, and used in different ways! The meanings of *dipolar* and *nonpolar* as used here are explained in the following material. (Many compilations adopt a different approach, based on polar and nonpolar denoting solvents having and not having a dipole moment. An example of a dipole moment is in the sketch below.)

methanol *dipolar protic*

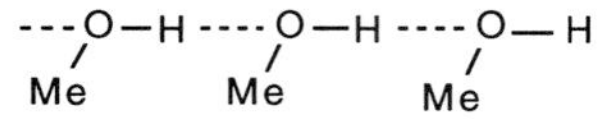

```
---O—H ----O—H ----O—H
  /        /        /
 Me       Me       Me
```

strong association by hydrogen bonding

benzene *nonpolar aprotic*

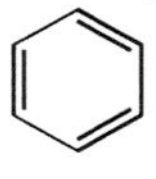

very little intermolecular association

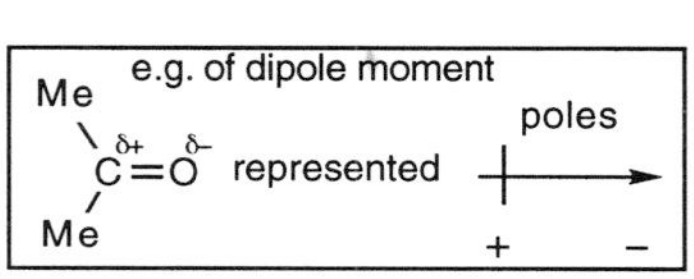

propanone *dipolar aprotic*

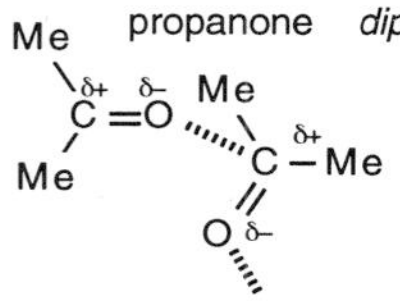

weak association by dipole–dipole attraction

e.g. of dipole moment

```
Me
 \ δ+  δ−                  poles
  C=O   represented  |——————▶
 /
Me                       +      −
```

Three types of solvents are shown; they differ in their propensity for intermolecular bonding. The influence of a solvent on an organic reaction depends mainly on two features: its *dielectric constant* and its *ability to solvate ions*.

The definition of dielectric constant (ε), a number, is not needed here: its significance is as follows. As a solvent's ε increases the electrostatic attraction between a solute's ions decreases. The stability of the ions therefore increases, and the solute has a greater tendency to ionise. ε is a macroscopic ('bulk') property, not concerned with details of the ions' state. Solvents with $\varepsilon > 20$ are termed *dipolar*, $\varepsilon < 20$ *nonpolar* (see **1.13Ge**). In general, dipolar solvents have relatively high dipole moments, nonpolar solvents relatively low ones. (Exceptions, e.g. pyridine, nonpolar, but dipole moment 2.37 debyes.)

In ability to solvate ions the main consideration is whether the solvent has an $H^{\delta+}$ which bonds to a negative ion. If so the solvent is termed *protic*, if not *aprotic*. The terms do *not* mean having an H, no matter of what sort, and devoid of an H. The scheme below shows a metal bromide in two solvents. An important feature is that many salts are soluble in dipolar aprotic solvents; their *anions* are not solvated and are therefore *reactive*.

Section 1.11 refers to Hal⁻ as a nucleophile in protic solvents which bond to Hal⁻. The extent of solvation is determined by the tendency of the different Hal⁻ to form H bonds, $I^- < Br^- < Cl^- < F^-$. The solvent cage impedes, and must be removed during, reaction as a nucleophile. Thus F⁻ which is most solvated has the lowest nucleophilicity etc. In aprotic solvents Hal⁻ is not solvated. Nucleophilicity is determined by the energy (enthalpy) of the C–Hal bond being formed. C–F is the strongest bond (see **2.1Ge2b**), hence F⁻ has the highest nucleophilicity etc. and the nucleophile order is $F^- > Cl^- > Br^- > I^-$.

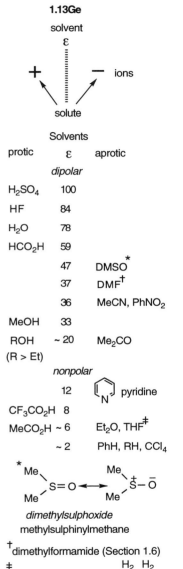

1.13Ge

solvent

ε

ions

solute

Solvents		
protic	ε	aprotic
	dipolar	
H_2SO_4	100	
HF	84	
H_2O	78	
HCO_2H	59	
	47	DMSO*
	37	DMF†
	36	MeCN, $PhNO_2$
MeOH	33	
ROH (R > Et)	~ 20	Me_2CO
	nonpolar	
	12	pyridine
CF_3CO_2H	8	
$MeCO_2H$	~ 6	Et_2O, THF‡
	~ 2	PhH, RH, CCl_4

* $\underset{Me}{\overset{Me}{\diagdown}}S=O \longleftrightarrow \underset{Me}{\overset{Me}{\diagdown}}\overset{+}{S}-\bar{O}$

dimethylsulphoxide
methylsulphinylmethane

† dimethylformamide (Section 1.6)

‡ *tetrahydrofuran* oxolan

$\underset{H_2C\diagdown_{O}\diagup CH_2}{\overset{H_2\;\;H_2}{\overset{C-C}{\diagup\;\;\;\diagdown}}}$

M⁺ Br⁻

in H₂O *dipolar protic*   in DMF *dipolar aprotic*
(several solvent molecules rather just one molecule, as shown, are involved in each case)

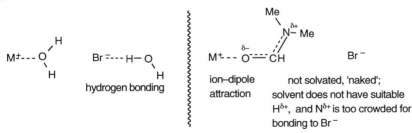

M±--- O⟨H,H⟩   Br =---H–O⟨H⟩   hydrogen bonding

M^+ -- $O^{\delta-}$ ≐ CH—$N^{\delta+}(Me)(Me)$ Br^-

ion–dipole attraction   not solvated, 'naked'; solvent does not have suitable $H^{\delta+}$, and $N^{\delta+}$ is too crowded for bonding to Br^-

2   Organic halides

Here, and where required in other chapters, the material is divided into sections covering the types within the group. Thus Section 2.1 refers to alkyl halides (saturated compounds), Sections 2.2 and 2.3 to halides containing a double bond. The order within each section is general tendencies (**Ge**), methods of preparation (**Pr**), and the characteristic reactions (**Re**).

Schemes such as **2.1Ge1**, **2.1Ge2** etc. are used throughout to present the chemistry. A scheme is usually followed by the related discussion but spacing requirements necessitate other layouts in some places. The discussions are brief; the meat of the subject is in the schemes themselves. At first sight some of the schemes appear rather daunting because they contain such a mass of information. *The best approach is to go through them slowly, writing out the material on scrap paper.* This will be helpful in understanding the chemistry and, later, in memorising the main points.

2.1   Alkyl halides

While the chlorides, bromides, and iodides fall into a graded series the fluorides are of a different nature. The monofluorides, which are very unreactive towards nucleophiles, are little used in general work and are not covered here. Scheme **2.1Ge1** shows the main features of alkyl halides.

2.1Ge1

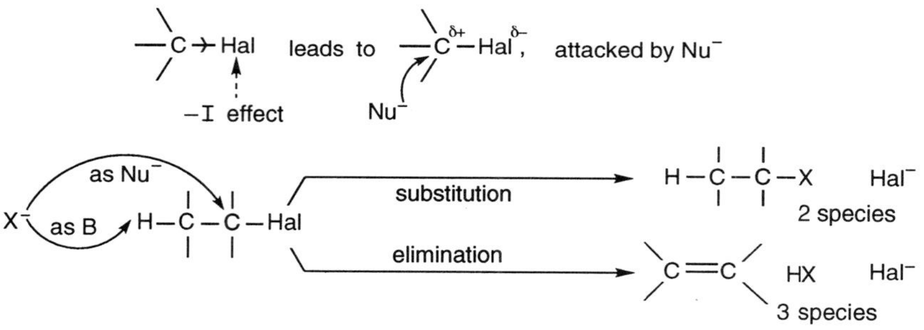

In the polarised halide Hal$^{\delta-}$ is relatively stable but C$^{\delta+}$ is unstable (hence the reactive site) and is attacked by nucleophiles. The importance of the halides lies mainly in their reactions with nucleophiles (**2.1Re1**, which comes later). The second general reaction to be considered is the attack of a base (B) on halides. However, before turning to these reactions we should deal with an awkward point which appears in different guises and often causes confusion.

The αH of the halide is more acidic than the βH, and so is removed more easily (margin). In general the carbanion formed merely equilibrates with the original halide; there is no 'outcome'. (#With R–CHHal$_2$ the carbanion may lead to dihalogenocarbenes, useful diradicals.) This and similar non–productive equilibria are justifiably omitted from textbooks, and no more will be shown here; attention will be directed to reactions that 'get somewhere'.

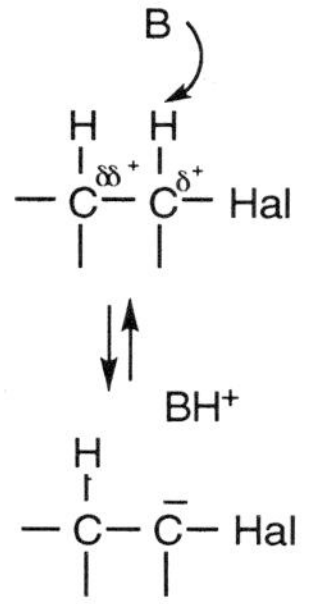

2.1Ge1 depicts a halide being attacked by X$^-$ acting as a nucleophile and as a base. X$^-$ may be negative, e.g. CN$^-$, or neutral, e.g. Me$_3$N. If X$^-$ acts as a nucleophile, the result is substitution. If X$^-$ acts as a base at the βH there is an important outcome, elimination giving an alkene. Consideration of the factors influencing the substitution/elimination ratio (**2.1Ge2**) illustrates the marked differences between primary, seconday, and tertiary halides.

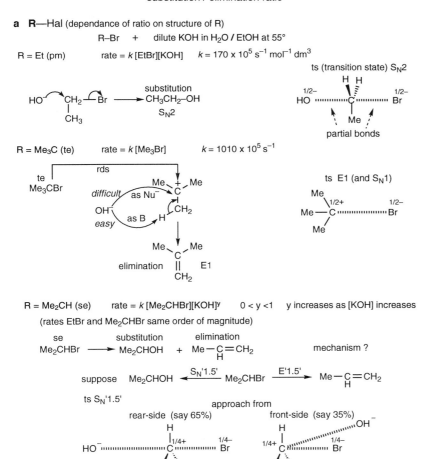

2.1Ge2

substitution / elimination ratio

a R—Hal (dependance of ratio on structure of R)

R–Br + dilute KOH in H$_2$O / EtOH at 55°

R = Et (pm)   rate = k [EtBr][KOH] k = 170 x 10^5 s^{-1} mol^{-1} dm^3

R = Me$_3$C (te)   rate = k [Me$_3$Br] k = 1010 x 10^5 s^{-1}

R = Me$_2$CH (se)   rate = k [Me$_2$CHBr][KOH]^y 0 < y <1   y increases as [KOH] increases

(rates EtBr and Me$_2$CHBr same order of magnitude)

a R

pm ⟶ substitution
se ⟶ substitution + elimination
te ⟶ elimination

substitution / elimination ratio decreases as pm ⟶ se ⟶ te

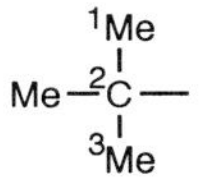

2-methylprop-2-yl (**4.1Ge**)
Structure and name are 'heavy', not convenient for everyday use!
Usually abbreviated to

t-butyl   Bu^t–
(tertiary)

Me$_3$C–Br = Bu^t–Br

Stereochemistry

(see chiral substates Section1.10)
S$_N$1   racemization
S$_N$2   inversion
assuming figures in ts sketch
'S$_N$1.5' 70% racemisation
            and 30% inversion

b Hal

substitution / elimination ratio

not markedly affected

b R—Hal

C—Hal bond energy (kJ mol^{-1})	C—F	C—Cl	C—Br	C—I
	485	339	284	209

rates of substitution and elimination increases as Cl $\longrightarrow$ Br $\longrightarrow$ I

c X$^-$

substitution / elimination ratio increases

as Nu$^-$ / B tendency of X$^-$

increases

c X$^-$

e.g. Me$_2$CH—Br $\xrightarrow{\text{Na}^+ \text{ }^-\text{OR}}$ Me$_2$CH—OR (ether) + Me—C(H)=CH$_2$ (alkene)

ether / alkene ratio increases as R goes Me$_3$C $\longrightarrow$ Me

see **1.12Ge** Na$^+$ $^-$OMe is weaker B but stronger Nu$^-$ than Na$^+$ $^-$OCMe$_3$

d Temperature

substitution / elimination ratio

decreases as temp increases

d Temperature

see general reactions in **2.1Ge1** in substitution 2 species $\longrightarrow$ 2 species

in elimination 2 species $\longrightarrow$ 3 species

as the temperature is raised the amount of the elimination product increases

$$\Delta G^{\ddagger} = \Delta H^{\ddagger} - T\Delta S^{\ddagger}$$

$\Delta G^{\ddagger}$ = free energy of activation, difference in free energy beween ts and reactants

$\Delta H^{\ddagger}$, $\Delta S^{\ddagger}$ are corresponding enthalpy and entropy of activation

$\Delta S^{\ddagger}$ bigger for elimination than for substitution

The influence of the alkyl group's structure is covered in part **a.** OH$^-$ can act as a strong nucleophile (**1.11Ge**) or as a strong base (**1.7Ge**). With EtBr there is little hindrance to rear–side approach at the αC, and the carbocation CH$_3$—CH$_2$$^+$ would be very unstable. Thus OH$^-$ acts as a nucleophile, S$_N$2 ensues and EtOH is formed in high yield. With Me$_3$CBr there is hindrance to approach of OH$^-$ but the carbocation Me$_3$C$^+$ is stable (Section 1.10). Formation of Me$_3$C$^+$ is helped by another factor as follows. In Me$_3$CBr the central C is sp^3 hybridised, tetrahedral, and the angle between groups is 109.5°; in Me$_3$C$^+$ the central C is sp^2, trigonal, and the angle between groups is 120° (Section 1.9). Thus, as Me$_3$CBr ionises, the Me groups move farther apart, and there is relief of strain.This applies generally to tertiary C going from tetrahedral to trigonal. The effect was described originally, and very aptly, as 'relief of back strain' but, sadly, this term is no longer used. Attack of OH$^-$ as a nucleophile on the central C of Me$_3$C$^+$ is unfavourable because it is difficult to get into the crowded centre, and the change from trigonal to tetrahedral increases strain (reversal of relief attending ionisation).There is little impedance to attack of OH$^-$ as a base at the βH (and there are nine of them) on the periphery (**1.12Ge**); thus E1 is favoured and gives the alkene in high yield. *The amount of Me$_3$COH, formed by S$_N$1, is less than 5%.*

Me$_2$CHBr, a secondary halide, forms comparable amounts of ether and alkene. There are two views about the mechanism

(i) Some molecules react by S$_N$1 + E1 and some by S$_N$2 + E2. Thus:

Rate = k$_1$[Me$_2$CHBr] + k$_2$[Me$_2$CHBr][KOH]

This accords with the experimental results. As [KOH] increases the importance of the S$_N$2 + E2 term increases.This is the simplest but probably not the correct explanation.

(ii) S$_N$1 + E1 and S$_N$2 + E2 are *extremes of ranges of mechanisms*. All the molecules react by the same mechanisms, one for substitution and one for elimination, which are intermediate between the extremes. The notion of '1.5' reactions shown in **2.1Ge2** is for illustration only; *it is not to be taken literally.* ([#]Full treatment, later courses, involves consideration of intimate and solvent–separated ion pairs.) The basis of the present simplistic version is the difference in degree of involvement of OH$^-$ in the transition state. In S_N1 + E1, OH$^-$ is not involved. In S_N2 + E2, OH$^-$ is fully involved. In S_N'1.5' + E'1.5', OH$^-$ is involved to some extent. The schematic transition state for S_N'1.5' is an attempt to depict this, and a corresponding transition state for E'1.5' can be envisaged.

The variation in rates with the nature of Hal, part **b**, is in line with the L$^-$ order, which parallels the acid strength of the conjugate acids(**1.11Ge**). The variation also follows the sequence of C–Hal bond energies.The figures show C–I to be the weakest, easiest to break; hence reactions of iodides are fastest. There is a possible caveat: the figures are energies for homolytic breaking (C–Hal giving C· and Hal·) not the heterolytic breaking involved here.

Varying the nature of X$^-$, part **c**, has the expected effect. The trend towards elimination with increasing temperature, part **d**, may arise from the entropy factor, which favours elimination (two species forming three) relative to substitution (two giving two). The equation in **2d** shows the thermodynamic relation. As temperature increases the free energy of activation is lowered more for elimination than for substitution; hence $k_{elimination}$ increases more than $k_{substitution}$.

The preparations are shown in detail in scheme **2.1Pr**.

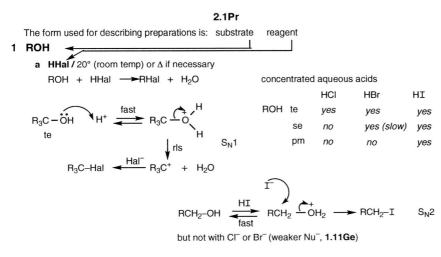

2.1Pr

The form used for describing preparations is: substrate   reagent

1 ROH

a HHal / 20° (room temp) or Δ if necessary

ROH + HHal ⟶ RHal + H$_2$O   concentrated aqueous acids

		HCl	HBr	HI
ROH	te	*yes*	*yes*	*yes*
	se	*no*	*yes (slow)*	*yes*
	pm	*no*	*no*	*yes*

but not with Cl$^-$ or Br$^-$ (weaker Nu$^-$, **1.11Ge**)

HHal may lead to rearrangement

Me$_3$C–CH$_2$–OH
  │ 'dry' HBr
Me$_2$C–CH$_2$—Me
  │
  Br

Think of protonation; loss of H$_2$O accompanied by migration of Me (as carbanion) to give more stable carbocation; addition of Br$^-$

Potency of HHal increased by using e.g.

'dry' HBr (no H$_2$O)   reaction potentially reversible, absence of H$_2$O drives reaction

'dry' HBr / sq H$_2$SO$_4$ increases amount of protonated alcohol   (sq = small quantity)

HCl / ZnCl$_2$ forms   RCH$_2$—Ö$^+$(H)/ZnCl$_2^-$ } better L$^-$ than H$_2$O

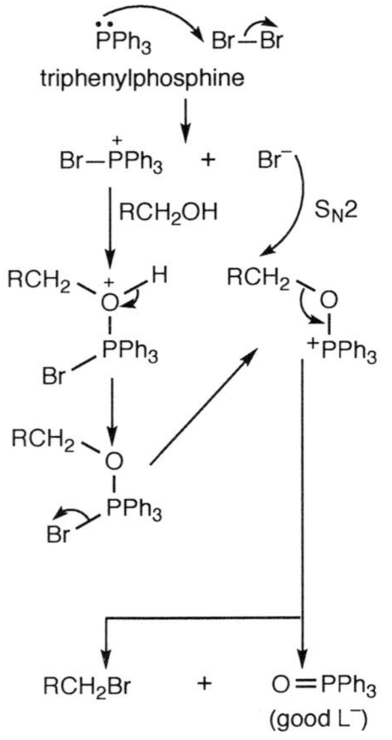

$\overset{..}{PPh_3}$ $Br \overset{\frown}{\underset{\smile}{}} Br$

triphenylphosphine

$\overset{+}{Br}-PPh_3$ + Br^-

RCH_2OH | S_N2

$RCH_2 \underset{Br}{\overset{+}{\underset{|}{\overset{O}{\underset{PPh_3}{|}}}}} \overset{H}{}$ $RCH_2 \underset{+PPh_3}{\overset{O}{}}$

$RCH_2 \underset{Br}{\overset{O}{\underset{|}{PPh_3}}}$

RCH_2Br + $O=PPh_3$
(good L^-)

no rearrangement
Me_3C-CH_2-OH
↓
Me_3C-CH_2-Br

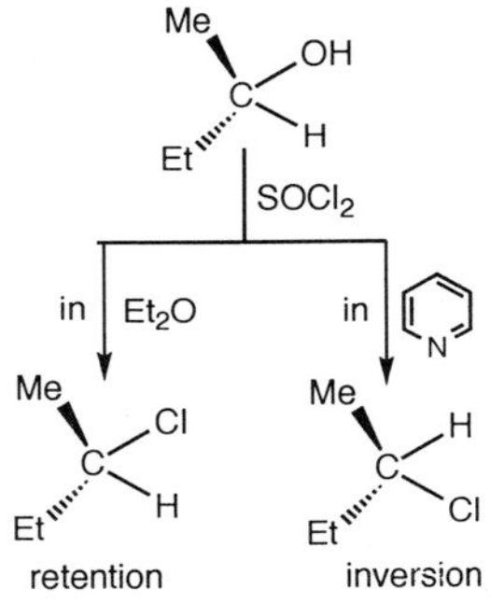

retention   inversion

b P–Hal reagents   driving force is formation of very strong PO double bond

ROH + $PHal_3$ ⟶ $RHal$ + $\overset{O}{\underset{H}{\diagdown}}PHal_2$

yields pm > se > te ◄---- tend to give elimination

$PHal_3$ (Cl, Br, I) or mixtures P/Br_2, P/I_2 $PBr_3 /$ [pyridine] / 10° is good method
several modern developments e.g. PPh_3/Br_2, PPh_3/CCl_4

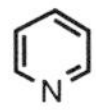

c SOCl₂ (thionyl chloride)

simple representation:

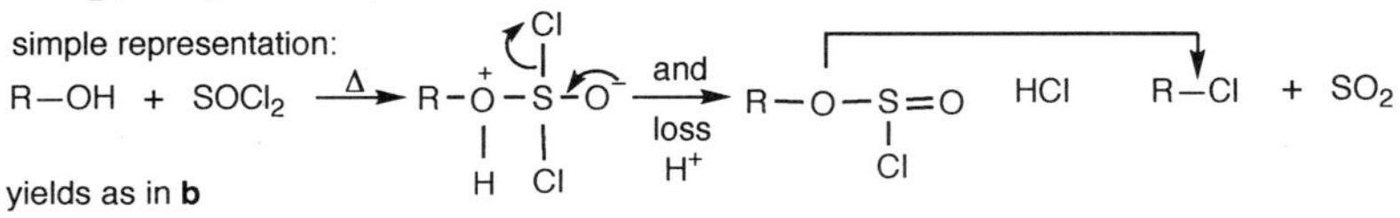

$R-OH$ + $SOCl_2$ $\xrightarrow{\Delta}$ $R-\overset{+}{\underset{H}{O}}-\overset{Cl}{\underset{Cl}{S}}\overset{\frown}{O}$ $\underset{H^+}{\overset{and}{\underset{loss}{}}}$ $R-O-S=O$ HCl $R-Cl$ + SO_2

yields as in **b**

#Et_2O as solvent, HCl formed is largely covalent, little Cl^- present

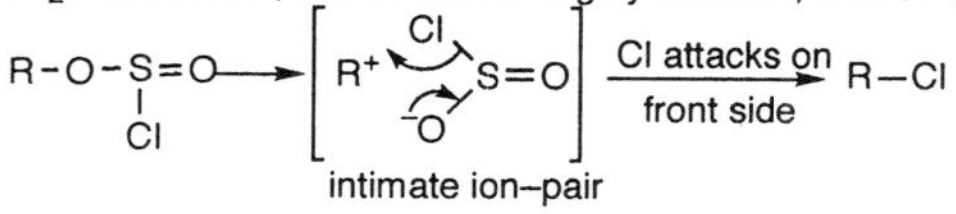

$R-O-S=O \longrightarrow \left[R^+ \overset{\frown}{\underset{O}{}}\overset{Cl}{S}=O \right] \xrightarrow[\text{front side}]{Cl\ attacks\ on} R-Cl$ $S_N i$
    $\underset{Cl}{|}$ intimate ion–pair   Substitution by an
                              internal nucleophile

pyridine as solvent, with HCl gives [pyridine $\overset{+}{N}H$] Cl^-, $Cl^- \overset{\frown}{R}-\overset{\frown}{O}-S=O \longrightarrow Cl-R$ S_N2
                                                          $\underset{Cl}{|}$ + SO_2 + HCl

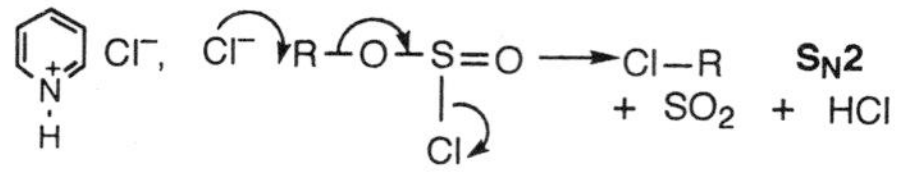

2 Hydrocarbons   4.2Re1   5Re1a,2a,4a,4b
(These are references forward to reactions shown and discussed in later chapters)

3 RHal (i.e. from halides obtained by the other methods given here)

RCH_2-Cl (or Br) + NaI $\xrightarrow{Me_2CO/\Delta}$ RCH_2-I + $NaCl$ (or Br) pm only **S_N2**
       soluble in Me_2CO much less soluble, precipates

direction contrary to expectation from nucleophilicities in aprotic solvents (Section 1.13);
reaction driven by precipitation of NaCl (or Br)

useful development: RCH_2-OH $\xrightarrow[\text{(Section 1.4)}]{TsCl/[pyridine]}$ RCH_2-OTs $\xrightarrow[DMF/\Delta]{NaI (or Br)/}$ RCH_2-I (or Br)

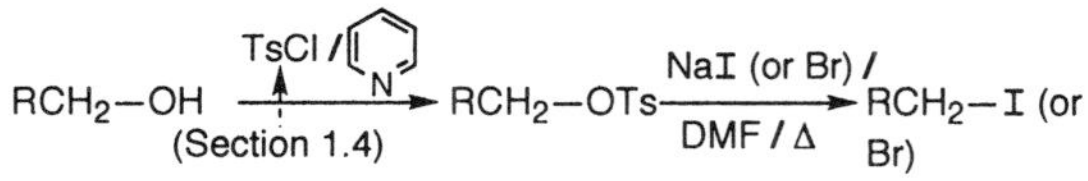

4 RCO₂H

$R-CO-OAg$ + Br_2 $\xrightarrow{CCl_4/\Delta}$ $R-Br$ + $AgBr$ + CO_2

Hunsdiecker reaction   #mechanism, radical chain reaction
overall effect: $R\!-\!\boxed{CO_2H}$ ⟶ $R\!-\!\boxed{Br}$ pm, se, te
no rearrangement $Me_3C-CO_2H \rightarrow Me_3C-Br$ $Me_3C-CH_2-CO_2H \rightarrow Me_3C-CH_2-Br$
convenient modification: use RCO_2H/HgO instead of RCO_2Ag

Protonation is essential in **2.1Pr1a**; the L^- is H_2O. Reaction of ROH itself would involve OH^- as L^-, a very unlikely event(**1.11Ge**).The first step is set out in detail here; from now on the curly arrows are omitted from such simple processes. High reactivity with tertiary ROH is as expected. The second step involves Hal^- acting as a nucleophile with R_3C^+. This appears to conflict with the trend of **2.1Ge2a**. However in protic solvents HHal are very strong acids (pK_a values **1.11Ge**), and Hal$^-$ are therefore very weak bases. Further, HHal is present in massive excess, so any alkene generated by attack of B (unlikely) would be converted back to the carbocation by protonation. Reaction of the primary ROH, lowest reactivity, is successful only with HI. With those alcohols which do not react with concentrated aqueous HHal various devices (in scheme) are used to pep up the reactivity.

The mechanism of a modern development of **Pr1b** is illustrated in the margin. No HBr is present, and substitution occurs without rearrangement. As noted earlier, the same alcohol with HBr gives a rearranged bromide.

The name of the chemist who invented the reaction is given under **Pr4**. There is no great virtue in remembering names, but they do give a convenient shorthand way of referring to reactions. Most of the names commemorate chemists, many German, who were responsible for developing the basis of the subject in the pioneering period *ca.* 1850–1910.

The reactions of RHal are shown in **2.1Re**. These will be encountered again in the chapters dealing with the products and, if necessary, discussed at the second meeting. That so many types of compound can be obtained justifies placing the halides at the centre of functional group chemistry.

2.1Re

1 With Nu$^-$ Yields pm—good, se—modest, te—poor Δ unless stated otherwise

a R–Hal $\xrightarrow[\text{EtOH}]{\text{KOH / H}_2\text{O /}}$ R–OH   *alcohols*   **b** R–Hal $\xrightarrow{\text{Na}^+\ ^-\text{O–R}^1}$ R–O–R^1 *ethers*

‑‑‑ to dissolve RHal   R and R$'$ may be identical or different; highest yields with both pm

c R–Hal $\xrightarrow[\text{Et}_2\text{O}]{\text{R}^1\text{–CO–OAg /}}$ R–O–CO–R^1 *esters*   **d** R–Hal $\xrightarrow{\text{K}^+\text{ SH}^-\text{ / EtOH}}$ R–SH   *thiols* (vile smell)

e R–Hal $\xrightarrow{\text{K}^+\ ^-\text{SR}^1\text{ / EtOH}}$ R–S–R^1 *thioethers* (vile smell)   **f** R–Hal $\xrightarrow{\text{H–N}\diagdown}$ R–N$\diagdown$ *amines*

e.g. Me–Br + NH$_3$ → MeNH$_2$

g R–Hal $\xrightarrow{\text{R}_3^1\text{N}}$ R–$\overset{+}{\text{N}}$R$_3^1$ Hal$^-$ *quaternary ammonium salts*   **h** R–Hal $\xrightarrow{\text{R}_3^1\text{P}}$ R–$\overset{+}{\text{P}}$R$_3^1$ Hal$^-$ *phosphonium salts*

i R–Hal $\xrightarrow[\text{(sodium azide)}]{\text{NaN}_3\text{ / MeOH}}$ R–N$_3$ *azides*

see notes about **i,j** and **k** at bottom of this scheme

j R–Hal $\xrightarrow{\text{NaNO}_2\text{ / DMF / 20°}}$
R–$\overset{+}{\text{N}}$$\overset{\overset{\text{O}}{\|}}{}$ *nitro compounds* O$^-$ major product

more convenient than
AgNO$_2$ / Et$_2$O

R–O–N=O *nitrite esters* minor product

k R–Hal $\xrightarrow[\text{EtOH or in DMSO}]{\text{NaCN in H}_2\text{O /}}$ R–C≡N *nitriles* ~90%

Et–I $\xrightarrow[\text{no solvent}]{\text{AgCN}}$ Et–$\overset{+}{\text{N}}$≡$\overset{-}{\text{C}}$ *isonitrile* (50%) (vile smell)

AgCN reaction is complex, reliable results not available for many R–Hal; not general Pr of R–NC

l R–Hal $\xrightarrow{^-\text{C}≡\text{C–R}^1}$ R–C≡C–R^1 *disubstituted alkynes*   **m** R–Hal $\xrightarrow{\text{CH}_3\text{–CO–}\overset{-}{\text{C}}\text{H–CO}_2\text{Et}}$ CH$_3$–CO–CH–CO$_2$Et

R (below CH)

n R–Hal $\xrightarrow{\ ^-CH(CO_2Et)_2\ }$ R–CH(CO_2Et)_2   **o** R–Hal $\xrightarrow{\ Li\bar{C}uR_2^1\ }$ R–R^1 **4.2Pr2c**

p R–Hal $\xrightarrow{\ ArH\ /\ AlCl_3\ }$ R–Ar   e.g. $\xrightarrow{\ Me_3C-Cl\ /\ }_{AlCl_3}$

Friedel–Crafts reaction, mechanism different from others here, range of R–Hal can be used

Re1i   azide ion N_3^- strong Nu$^-$(**1.11Ge**), linear (like a slim pencil), so very small if viewed from end   alkyl azides

$\bar{N}=\overset{+}{N}=\bar{N}$ ⟵------ ⊬⟩ $R-\bar{N}-\overset{+}{N}\equiv N \longleftrightarrow R-N=\overset{+}{N}=\bar{N}$

Re1j,k NO_2^- and $C\equiv N^-$ are examples of ambident ('either tooth') anions. Generalisation: increasing δ+ of C in transition state favours attack by more negative centre (O > N, N > C)

In **1k** pull by Ag^+ increases positive on C of bromide; favours formation of isonitrile $R-\overset{\delta++}{CH_2}-Br\text{---}Ag^+$

2 With B

Elimination of HHal (dehydrohalogenation), tendency with RHal te > se > pm, but high yields can be obtained with all using appropriate B

See **5Ge2** for discussion of elimination

e.g.

P ⟵------------ main products ------------⟶ **Q**   probably E2 in nonpolar aprotic solvent

A cautionary note. Although often find stated that $CH_3-CH_2-Br \xrightarrow{\ KOH\ /\ EtOH\ /\ \Delta\ } H_2C=CH_2$ the yield of ethene is 0.9%

In fact $HO^- + EtOH \rightleftharpoons EtO^- + H_2O$ Et–O–Et   *main* product

3 Influence of reagents and conditions on behaviour of te RHal

a $Me_3C-Br \xrightarrow{\ KCN\ } Me-\underset{H}{C}=CH_2 + Me_3C-CN$ [to prepare this nitrile start from Me_3C-CO_2H]

        main product   < 5%

CN^- very strong Nu$^-$ (**1.11Ge**), medium weak B (**1.7Ge**), but with te RHal tendency for elimination so strong that CN^- acts as B

b Problem: how to convert Me_3C-Br directly into Me_3C-OH ?

   Solution: $Me_3C-Br \xrightarrow[\text{stir at } 20° \text{ for several hours}]{\ AgNO_3\ /\ H_2O\ (\text{excess})\ /\ } Me_3C-OH$ yield > 80%   mechanism ? S_N'1.5'

c $\xrightarrow{\ LiN_3\ /\ Me_2CO\ }$ S_N2

4 Reduction

a R–Hal $\xrightarrow{\text{LiAlH}_4 / \text{Et}_2\text{O}}$ R–H $\xleftarrow{\text{NaBH}_4 / \text{DMSO}}$ R–OTs $\longleftarrow$ R–OH

best pm Br or I   pm and se

M = Al or B   L⁻ = Hal⁻ or TsO⁻

b R–Hal $\xrightarrow[\text{C}_6\text{H}_6 / \Delta]{\text{Bu}_3\text{SnH} / \text{sq radical initiator} /}$ R–H   tributylstannane

mechanism for later reference (see **4.2Re1a** for representation of radical reactions)

overall, chain reaction:

Bu₃Sn–H ⟶ R–Hal

Bu₃Sn–Hal ⟵ R–H

radical initiator e.g. Me₂C–N=N–CMe₂ $\xrightarrow{\text{warm}}$ 2 Me₂Ċ–CN + N₂

| | |
CN CN

initiation: Me₂Ċ–CN + H–SnBu₃ ⟶ Me₂CH–CN + ṠnBu₃

propagation: Bu₃Ṡn + Hal–R ⟶ Bu₃SnHal + Ṙ   Ṙ + H–SnBu₃ ⟶ ṠnBu₃ + R–H

chain carrier

5 With M (metal)   3Pr

2.1Re1a—o consists of a range of convenient efficient reactions which provide access to many other functional groups; as already noted, discussion of them is deferred. Although the general form R–Hal is used for the substrate the most effective halides are primary bromides, RCH₂Br. Primary iodides, RCH₂I, are more expensive and sometimes give side–reactions which lower the yield of the required product. The solvents shown are widely used but others are equally good in most cases. (Don't try to remember the details.)

2.1Re2 illustrates a general point: with some reactions the outcome can be changed in the desired direction by adroit choice of reagent. In the example selected product **P** is more stable than **Q** (**5Ge1**); other things being equal **P** would be formed. However, approach of the base to the CH₃ (leading to **Q**) involves less steric hindrance than approach to a ring CH₂ group (leading to **P**). Thus with a very big base (such as Et₃C–O⁻) attack occurs at the CH₃ not at a CH₂. The two reactions are *regioselctive*. (A regioselective reaction is one which may give two or more structural isomers but leads, exclusively or predominantly, to only one isomer.)

The outcome of **Re3a** is, as expected, elimination. In **3b** a tertiary halide is induced to undergo substitution. The key is in the conditions. No base is present (other than H₂O); the tendency for elimination is thus reduced. The temperature is low, which favours substitution (**2.1Ge2d**). Without Ag⁺ the reaction would be slow; Ag⁺ coordinates with the Br atom, thereby increasing the solubility of Me₃CBr, and then pulls it off as Br⁻. **3c** is remarkable, S_N2 with a tertiary halide. There are three contributing factors. In 1–bromo–1–methylcyclopentane the βCH₂ groups are held back from the αC–Br, thus reducing hindrance to attack at the C. N₃⁻ is a strong 'slim' nucleophile but a weak base. (See bottom of **2.1Re1** and pK_a of HN₃ in **1.11Ge**.) Me₂CO is an aprotic solvent of lowish ε (**1.13Ge**) and therefore inimical to formation of the ions involved in E1 and S_N1. The conclusion to be drawn is that reversal of general behaviour is rare and requires exceptional circumstances.

Re4a involves the very useful reagents LiAlH$_4$ (lithium aluminium hydride) and NaBH$_4$ (sodium borohydride). Using OTs as an alternative to Hal is convenient and, in several instances, advantageous. Tributylstannane(**4b**) is a selective reagent for replacing Hal by H.

2.2 Allyl halides

These have a CC double bond attached to the C–Hal as shown in **2.2Pr**.

2.2Ge

allyl **carbocation, carbanion,** and **radical** are stabilised by delocalisation (Section 1.6)

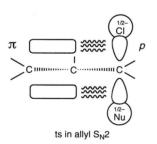

benzyl systems are similarly stabilised

canonicals with charge or unpaired electron at positions 2 and 6

ts in allyl S$_N$2

2.2Pr

general structure

parent allyl group $H_2C=C-C-$
                              H H$_2$

> The conventional representations of allyl bromide and vinyl bromide (see later) are
>
> $CH_2=CH-CH_2-Br$ and $CH_2=CH-Br$
>
> It is very tedious to draw these with the computer program being used; the program likes to produce the unusual but more correct forms adopted here

reagents involving Hal˙

e.g. $H_2C=C-CH_3$ $\xrightarrow{Cl_2\,/\,500°}$ $H_2C=\overset{1}{C}-\overset{2}{C}-\overset{3}{Cl}$ *allyl chloride*
         H   H H$_2$

252 kJmol^{-1}

3–chloroprop–1–ene

see also **5Re4b**

[Don't worry about names and numbering at this stage]

2.2Re

$H_2C=C-C-Cl$ $\xrightarrow[\textit{very fast}]{KOH\,/\,H_2O}$ $H_2C=\overset{3}{C}-\overset{2}{C}-\overset{1}{OH}$ *allyl alcohol*   S$_N$1
       H H$_2$ H H$_2$ prop–2–en–1–ol

via ↓   OH$^-$

$H_2C=C-CH_2^+\longleftrightarrow H_2C^+-C=CH_2$ $\left[H_2C\text{-----}\underset{H}{C}\text{-----}CH_2\right]^+$
       H   H   composite representation of carbocation

$H_2C=C-CH-CH_3$ $\xrightarrow{NMe_3}$ $H_2C=C-CH-CH_3$ + $H_2C-C=C-CH_3$
       H | H | | H H
          Cl $^+$NMe$_3$ Cl$^-$ $^+$NMe$_3$ Cl$^-$

The important feature is the very high reactivity towards nucleophiles. This stems from the enhanced stability of the allyl carbocation (**2.2Ge** in margin). Thus, by following the S$_N$1 path allyl halides undergo substitution reactions in which the rate-determining step is accelerated. The low C–Cl energy reflects this high reactivity. Even the chlorides are so reactive that the more reactive halides (Br, I) are seldom required. Benzyl halides (PhCH$_2$–Hal) are similarly reactive. With allyl and benzyl systems attention is usually confined to S$_N$1 reactions, which are readily explained. However the S$_N$2 rates also show acceleration. For chlorides, relative rates are: propyl 1, allyl ~100, benzyl ~300. [#]Of many explanations the following is probably the best. The transition state in S$_N$2 is stabilised by favourable interaction between the π elections and the proximate *p* orbital as indicated at the bottom of **2.2Ge**. With substituted allyl halides two products are usually formed (**2.2Re**), by reaction at either end of the unsymmetrical carbocation. This is *not* to be construed as reaction with different canonicals, which are hypothetical, not real, species.

2.3 Vinyl halides

These have the Hal directly attached to a CC double bond as shown in **2.3Ge**

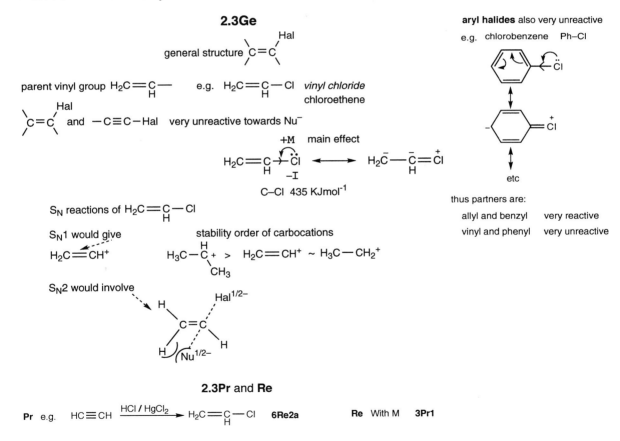

2.3Ge

general structure

parent vinyl group H$_2$C=C— e.g. H$_2$C=C—Cl *vinyl chloride* chloroethene

and —C≡C—Hal very unreactive towards Nu⁻

+M main effect

C–Cl 435 KJmol⁻¹

S$_N$ reactions of H$_2$C=C—Cl

S$_N$1 would give
H$_2$C=CH⁺

stability order of carbocations
H$_3$C—C⁺(H)(CH$_3$) > H$_2$C=CH⁺ ~ H$_3$C—CH$_2$⁺

S$_N$2 would involve

2.3Pr and Re

Pr e.g. HC≡CH —HCl / HgCl$_2$→ H$_2$C=C—Cl **6Re2a** **Re** With M **3Pr1**

aryl halides also very unreactive
e.g. chlorobenzene Ph–Cl

etc

thus partners are:

allyl and benzyl   very reactive
vinyl and phenyl   very unreactive

The important feature is the very low reactivity towards nucleophiles. (Vinyl chloride is shown in **2.3Ge** and used for discussion; vinyl bromide and iodide are slightly more reactive but not markedly so.) S$_N$1 and S$_N$2 reactions do not occur, or occur very slowly. Several features are involved. As C goes from sp^3 to sp^2 to sp the C becomes more electron attracting (**5Ge1**). Thus relative to ethyl the vinyl group is weakly electron attracting (**1.6Ge**). This opposes the inductive (**-I**) effect of Cl; the main effect of Cl is therefore mesomeric (**+M**). Appreciable contribution of the resulting charged canonical is confirmed by the high C–Cl energy. The C–Cl bond has partial double character, and to break it by either S_N mechanism is therefore difficult. S_N1 would lead to an unstable carbocation, but this is not so unstable as was once thought. More important is that S_N1 involves disruption of a stabilised molecule. In S_N2 the transition state must have a linear Nu—C—L arrangement (**2.1Ge2**). For vinyl systems this would involve some hindrance between the incoming nucleophile and a substituent (even H) on the βC (sketch in **2.1Ge2**). Despite their general unreactivity towards nucleophiles vinyl halides polymerise readily and give important industrial materials.

The important general relationships of **2.2** and **2.3** are the *high reactivity of allyl and benzyl halides* and the *low reactivity of vinyl and aryl halides*.

3   Organometallic compounds

This chapter covers the most useful organometallic compounds R–M with M = Li, Mg or Cd. Although Zn compounds have been largely superseded, one of their reactions still merits inclusion here. Lithium cuprates are treated only very briefly. The dialkyl derivatives of magnesium, R–Mg–R, are known, but the Grignard reagents, R–Mg–Hal, are the ones commonly used. In this chapter R–M stands for R–M compound(s) or reagent(s), thus obviating repetition of 'compound(s)' or 'reagent(s)'. Scheme **3Ge** portrays polarisation of the covalent R–M bond leading to a $C^{\delta-}$ centre; this dominates the chemistry of R–M and is responsible for tendencies **1** and **2**. Tendency **3**, a nuisance, is not so readily predicted.

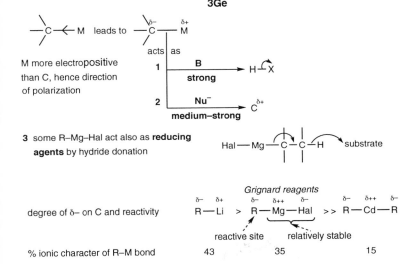

R–M are decomposed by water (see **3Re1**); they are also sensitive, in the order Li > Mg > Cd, to atmospheric moisture and oxygen. The general procedure is to generate R–M in an inert solvent such as Et_2O (*ether*, ethoxyethane], hexane, or benzene, and then to add the reagent to this solution. In a few cases the R–M solution is added to the reagent. Thus R–M are not isolated unless required for studies such as X–ray diffraction. R–Li are prepared and used under N_2. Solutions of the less sensitive R–Mg–Hal in ether, b.p. 35°, are sufficiently protected from atmospheric moisture and oxygen by the layer of ether vapour above the solution. With R–Cd–R the only requirement is exclusion of water. R–Li have many advantages over the older R–Mg–Hal and solutions of R–Li in hexane may be bought. This is a great convenience to busy chemists but an expensive luxury in large scale work. For comparable reactions with, say, 50g of Me_2CO the costs would be ~£60 for Bu–Li supplied commercially but only ~£7 for Bu–Mg–Br generated in the laboratory. (A historical note: Viktor Grignard, 1871–1935, developed the synthetic uses of R–Mg–Hal in his Ph.D. research at the University of Nancy. The outstanding importance of this work led to the well merited award of a Nobel Prize in 1912. However the reagents were discovered by Philippe Barbier, Grignard's supervisor, who suggested the study of their reactions. There was no dispute about 'priority'; with commendable generosity of spirit Barbier insisted that the reagents should bear only Grignard's name.)

The structures of a few solid R–M have been investigated by X–ray diffraction. In solutions there are equilibria between various species, and the finer details remain obscure. Some of the results are summarised in scheme **3Ge.** For our purposes the simple representations R–M and R–Mg–Hal are adequate.

The wide range of general reactions exhibited by R–M put them in the first rank of reagents used in synthesis. They are the $C^{\delta-}$ complement of the $C^{\delta+}$ in organic halides; the two groups have central positions in functional group chemistry. Although most of the reactions (in **3Re**, which comes after the preparations) are depicted as occurring with R–Mg–Hal, the treatment is intended to include all the R–M covered here. In general, R–Mg–Hal and R–Li are similar; many of the reactions discussed later occur with both. There are, however, important differences between them, and these are clearly identified. The Cd and Zn reagents, which differ markedly from R–Mg–Hal, have particular applications and these are slotted in at appropriate places. ([#]The lithium cuprates, $LiCuR_2$, are very useful in certain reactions where R–Mg–Hal are unsatisfactory. The preparation is shown here, **3Pr2a(ii)**, and an application is given in **4.2Pr2e**. Full treatment, later course on synthesis.)

Scheme **3Pr** covers the two main methods for preparing R–M. All types of R–Hal (primary, secondary, and tertiary; Cl, Br, I) react with Li and Mg (**Pr1**). With primary and secondary Cl the reactions are inconveniently slow; with tertiary Br they are very fast and difficult to control. The halides generally used are shown in the scheme. MeI, a liquid b.p.42°, is easier to handle than MeBr, a gas b.p.3°. Allyl and benzyl halides are so reactive that they undergo S_N2 reactions with the R–M being formed,

i.e. R–Hal + R–M $\longrightarrow$ R–R + M–Hal.

Conditions for reducing this tendency are shown in **Pr1**. Et₂O or, in a few cases, THF (tetrahydrofuran) are the solvents used with R–Mg–Hal. R–Li react with Et₂O at 20°; Et₂O may be employed at –70° but hexane, which is inert, is generally preferred. Benzene is the standard solvent for R–Cd–R.

3Pr

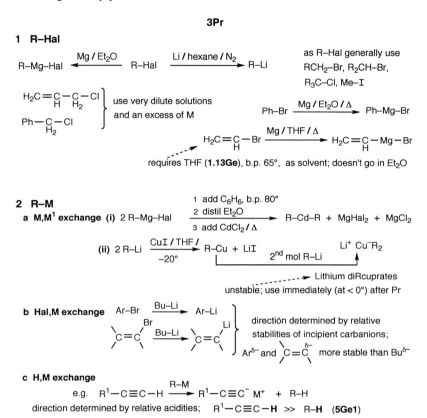

1 R–Hal

R–Mg–Hal ←⎯ Mg / Et₂O ⎯ R–Hal ⎯ Li / hexane / N₂ ⎯→ R–Li

as R–Hal generally use RCH₂–Br, R₂CH–Br, R₃C–Cl, Me–I

H₂C=C(H)–C(H₂)–Cl
Ph–C(H₂)–Cl
} use very dilute solutions and an excess of M

Ph–Br ⎯ Mg / Et₂O / Δ ⎯→ Ph–Mg–Br

H₂C=C(H)–Br ⎯ Mg / THF / Δ ⎯→ H₂C=C(H)–Mg–Br

requires THF (**1.13Ge**), b.p. 65°, as solvent; doesn't go in Et₂O

2 R–M

a M,M¹ exchange (i) 2 R–Mg–Hal ⎯ 1 add C₆H₆, b.p. 80° / 2 distil Et₂O / 3 add CdCl₂ / Δ ⎯→ R–Cd–R + MgHal₂ + MgCl₂

(ii) 2 R–Li ⎯ CuI / THF / –20° ⎯→ R–Cu + LiI ⎯ 2nd mol R–Li ⎯→ Li⁺ Cu⁻R₂ → Lithium diRcuprates unstable; use immediately (at < 0°) after Pr

b Hal,M exchange Ar–Br ⎯ Bu–Li ⎯→ Ar–Li

C=C(Br) ⎯ Bu–Li ⎯→ C=C(Li)

} direction determined by relative stabilities of incipient carbanions; Ar^δ⁻ and C=C^δ⁻ more stable than Bu^δ⁻

c H,M exchange

e.g. R¹–C≡C–H ⎯ R–M ⎯→ R¹–C≡C⁻ M⁺ + R–H

direction determined by relative acidities; R¹–C≡C–H ≫ R–H (**5Ge1**)

R–M prepared by **Pr1** are sources of other R–M which cannot be generated efficiently from M + R–Hal. Several variations of this approach are set out in **Pr2**. As the electropositivity of a metal increases so does its tendency to assume the ionic state. This determines the direction of **2a**. In **2a(i)** Mg is more electropositive than Cd; in **2a(ii)** Li is more electropositive than Cu. The outcome is that a more reactive R–M generates a less reactive R–M. Features responsible for the directions of **2b** and **2c** are shown in the scheme. Some syntheses would be difficult if it was not possible to form R–M from vinyl and phenyl halides. Despite the unreactivity of these halides in S_N reactions (**2.3Ge**) the Mg and Li derivates of vinyl and aryl systems can be obtained as shown in **Pr1** and **2**.

Scheme **3Re** deals with the reactions of R–M.

3Re

For convenience the representation $\overset{\delta-}{R}-\overset{\delta+}{MgHal}$ is used rather than the more elaborate $\overset{\delta-}{R}-\overset{\delta++}{Mg}-\overset{\delta-}{Hal}$ of **3Ge**

1 As B

$$\overset{\delta-}{R}-\overset{\delta+}{MgHal}$$
$$\overset{\delta+}{H}-\overset{\delta-}{X}$$
$$\longrightarrow \quad \begin{matrix} R \\ | \\ H \end{matrix} + \begin{matrix} MgHal \\ | \\ X \end{matrix}$$

H—X = H—O—   H—S—   H—N⟨   H—C≡C—   but not   H—CH$_2$—Me

↑

$pK_a \sim 45$

2 As Nu⁻

a with aldehydes and ketones: formation of *alcohols*

$$\overset{\delta+}{C}=\overset{\delta-}{O} \xrightarrow{\quad R-MgHal \quad} \begin{matrix} \quad O-MgHal \\ C \\ \quad R \end{matrix} \xrightarrow[\text{or} \atop NH_4Cl/H_2O]{\text{dilute } H_2SO_4} \begin{matrix} \quad O-H \\ C \\ \quad R \end{matrix} + Mg^{++} \text{ salts}$$

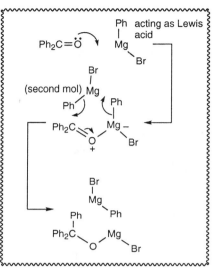

Ph acting as Lewis acid

R—CH$_2$—OH pm ⟵ R—MgHal ⟶

$$\begin{matrix} H \\ C=O \\ H \end{matrix}$$

$$\begin{matrix} R^1 \\ C=O \\ H \end{matrix}$$

$$\begin{matrix} R \\ CH-OH \\ R^1 \end{matrix} \text{ se}$$

$$\downarrow \begin{matrix} R^1 \\ C=O \\ R^2 \end{matrix}$$

$$\begin{matrix} R^1 \\ | \\ R^2-C-OH \\ | \\ R \end{matrix} \text{ te}$$

b with esters: formation of *alcohols*

$$R-MgHal \quad \begin{matrix} R^1 \\ \diagdown \quad O \\ C \\ \diagup \quad \diagdown O \\ OR^2 \end{matrix} \xrightarrow{1} \begin{matrix} R^1 \\ \diagdown \quad O-MgHal \\ C \\ \diagup \quad \diagdown \\ R \quad OR^2 \end{matrix} \xrightarrow{2} \begin{matrix} R^1 \\ \diagdown \\ C=O \\ \diagup \\ R \end{matrix} + R^2O-MgHal \qquad \begin{matrix} R^1 \\ | \\ R-C-OH \\ | \\ R \end{matrix} \text{ te}$$

$\boxed{k_3 > k_2 \sim k_1}$ hence $H-\overset{\overset{O}{\|}}{C}-OR^2 \longrightarrow H-\overset{\overset{R}{|}}{\underset{|}{C}}-OH$ se

$R^1-\overset{\overset{O}{\|}}{C}-OR^2$

| 3 | R—MgHal (then work-up) |

OR^2 is lost: any ester of methanoic acid gives a secondary ROH; any ester of a higher acid gives a tertiary ROH

c with oxiran (ethylene oxide): formation of *alcohols containing two more C atoms than R—MgHal*

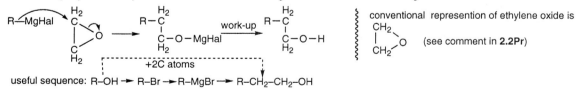

conventional represention of ethylene oxide is

$$\begin{matrix} CH_2 \\ | \quad \diagdown O \\ CH_2 \diagup \end{matrix} \quad \text{(see comment in 2.2Pr)}$$

useful sequence: R–OH → R–Br → R–MgBr → R–CH$_2$–CH$_2$–OH

d—h: methods using R–M to form *aldehydes and ketones*

d

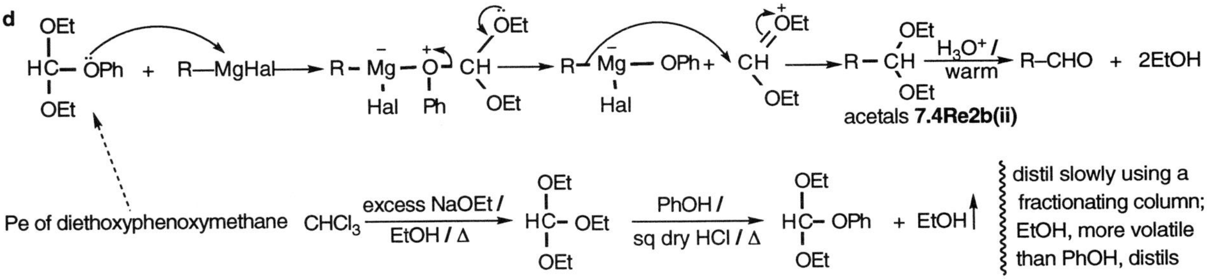

Pe of diethoxyphenoxymethane CHCl₃

$$\text{CHCl}_3 \xrightarrow[\text{EtOH}/\Delta]{\text{excess NaOEt}/} \underset{\text{OEt}}{\overset{\text{OEt}}{\text{HC–OEt}}} \xrightarrow[\text{sq dry HCl}/\Delta]{\text{PhOH}/} \underset{\text{OEt}}{\overset{\text{OEt}}{\text{HC–OPh}}} + \text{EtOH}\uparrow$$

distil slowly using a fractionating column; EtOH, more volatile than PhOH, distils

e

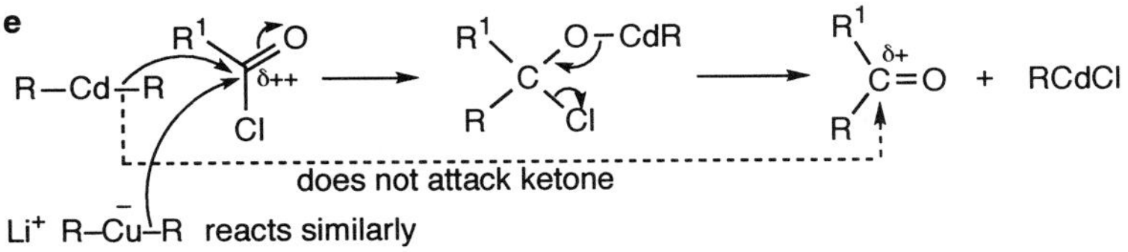

does not attack ketone

Li⁺ R–Cu–R reacts similarly

f R—MgHal (1 mol)

add slowly to a stirred solution of R¹–CO–Cl (1 mol) in THF at –70°

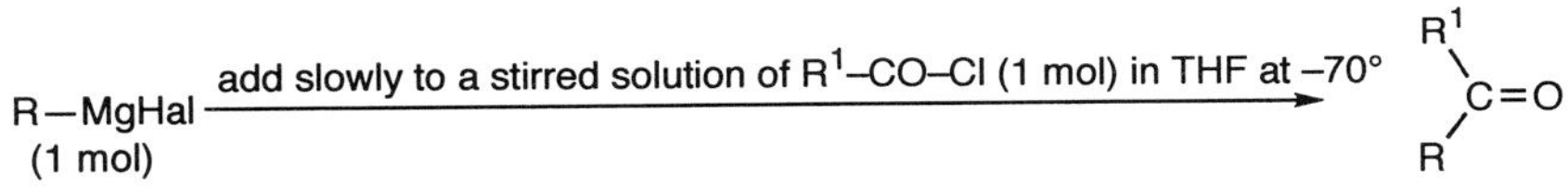

g

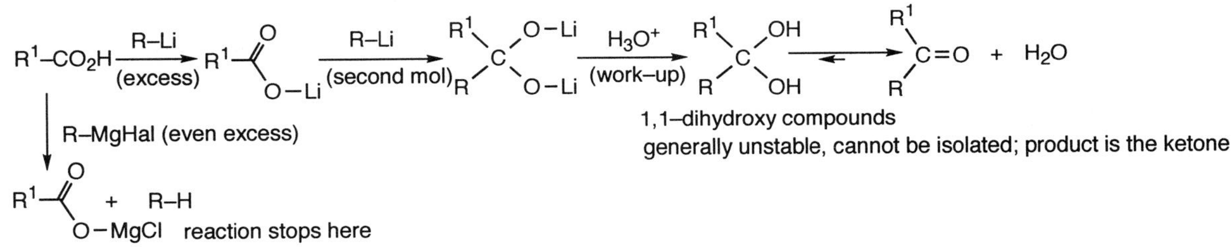

1,1-dihydroxy compounds generally unstable, cannot be isolated; product is the ketone

R–MgHal (even excess)

R¹–CO–O–MgCl + R–H reaction stops here

#h

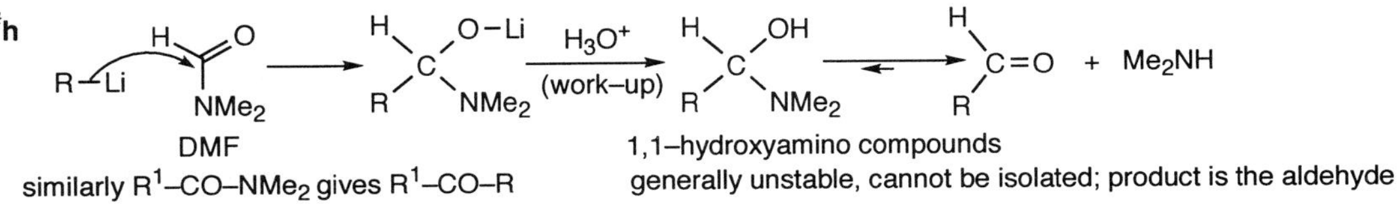

DMF

similarly R¹–CO–NMe₂ gives R¹–CO–R

1,1-hydroxyamino compounds generally unstable, cannot be isolated; product is the aldehyde

i with CO₂: formation of *acids*

R–MgHal   O=C=O   — add solid CO₂ to Et₂O solution →   O=C(R)(O–MgHal)   — work–up →   R–CO₂H

j use of Zn compounds: formation of *1,3–hydroxyesters*

R(R¹)C=O   — Zn / Br–CH₂–CO₂Et / C₆H₆ / Δ →   R(R¹)C(O–ZnBr)(CH₂–CO₂Et)   — work–up →   R(R¹)C(OH)(CH₂–CO₂Et)

Reformatsky reaction

structure of intermediate probably

BrZn–CH₂CO₂Et   ⟍   R(R¹)C=O   or   H₂C=C(OEt)(O–ZnBr)

3 R–MgHal as reducing agent: advantage of R–Li

R = Me₂CH

R₂C=O   — 1 R–MgBr (excess) / Et₂O / Δ for 3 hours ; 2 work–up →

5% R₂C(OH)R **U**   >85% ← [1 R–Li ; 2 work–up] R₂C=O

50% R₂C(OH)H **V**

45% R(R)C=O **W**

(R)(R)C=O ··· R–MgBr   **U**

R(R)C=O   H–MgBr   H–C–C–Me → R(R)C(O–MgBr)(H) + H₂C=CHMe   **V**

Me₂C–C(=O)–R   H   R–MgBr → Me₂C=C(O–MgBr)(R)   **W**

In **3Re1** the 'free bonds' go to C or H (Section 1.4). Thus H–O– denotes H–O–H and H–O–R. Although R-Mg-Hal are strong bases they do not abstract a proton from simple alkanes.

Re2 illustrate the main features of R–Mg–Hal as reagents in synthetic work. These reactions produce, initially, Mg derivatives from which the products are liberated by 'work–up', usually treatment with dilute acid. For products sensitive to acid, e.g. some tertiary ROH (see later), aqueous ammonium chloride is used. In representations of R–Mg–Hal reactions this work–up is generally omitted; it is regarded as included in the R–Mg–Hal specified. The simple mechanisms of the reactions in **Re2** are satisfactory for our purposes, but in reality the reactions may be more complicated. For example, certain cases of **2a** are known to proceed *via* a 6–membered transition state as shown. Reaction with esters(**2b**) gives ketones (or aldehydes) but these cannot be isolated under standard conditions; they react further to form alcohols, as explained in the scheme. The ethers Et₂O and THF (5–membered ring, not strained) solvate and thereby stabilise R–Mg–Hal. By contrast, oxiran (3–membered ring, strained) reacts smoothly with R–Mg–Hal (**2c**), the strain being relieved by ring opening. This is the crucial step in a sequence for adding 2C atoms to a substrate. **2d–h** are methods devised to circumvent the difficulty, already encountered(**2c**), of producing aldehydes and ketones. In **2d** the first stage gives an acetal (**10Re2**) which does not react with R–Mg–Hal. The mechanism of this reaction has not been established; that suggested here is speculative. The aldehyde is generated in a second separate stage, i.e. in the absence of any R–Mg–Hal. A different approach is adopted in **2e**. R_2Cd and $LiCuR_2$ are much weaker nucleophiles than R–Mg–Hal; R^1–CO–Cl, an acid chloride, is more reactive than a ketone towards nucleophiles (**8.1Ge1**). The upshot is that the Cd and Cu reagents do react with acid chlorides but not with ketones. Reaction **2f** calls for good experimental technique. The key features are that the R–Mg–Hal is, unusually, added slowly *to* the other reactant, and at low temperature. At –70° the difference in reactivity between R^1–CO–Cl and R–CO–R^1 is enhanced (see discussion of **4.2Re1c**), and the reverse addition ensures that the concentration of R–Mg–Hal at any moment is low. The divergence between R–Mg–Hal and R–Li in **1g** exemplifies the general higher reactivity of R–Li, and R–Li is more effective than R–Mg–Hal in **1h**. (Prefixing the intermediates in **1g** and **1h** by 1,1– means merely that the groups are attached to the same C.) The instability of these intermediates, which is not readily predicted by organic theory, stems from the decrease of free energy associated with their decomposition. (It does not follow that the reactions must be fast. Rates depend on activation energies, as illustrated in Section 1.5.) Both **1g** and **1h** embody the principle of **1d**, that the aldehyde or ketone is not generated in the presence of R–Mg–Hal. Reaction **2i** provides an excellent preparation of carboxylic acids(**8.2Pr1**).

A limitation of both R–Mg–Hal and R–Li is that the halide used in the preparation must not contain, elsewhere in the molecule, groups which react with R–M such as CO double bonds, CN double and triple bonds, carboxylic and nitro groups. 1,1–,1,2–, and 1,3–dihalides are also excluded: they undergo eliminations with Mg and Li. The use of other R–M may circumvent this difficulty. For example Zn organometallics (slightly more reactive than the Cd compounds) do not react with esters, and an ester group may be present in the reagent as shown in **2j**.

Reactions of R–Mg–Hal as a nucleophile are blown off course by even moderate steric hindrance. In **Re3** formation of product **U** requires approach of two C centres. Formation of **V** and **W** involving approach of a C centre and a H atom are less sterically demanding (see **1.12Ge**) and are therefore favoured. The R_2CO in the mixture of products does *not* arise from incompleteness of the R–Mg–Hal reaction. It is liberated in work–up from the Mg enolate shown in the scheme. Thus, **Re3** illustrates R–Mg–Hal acting as a nucleophile, a reducing agent, and a base. R–Li are effectively smaller, possibly because they are not solvated (in hexane). The formation of the tertiary alcohol in high yield by R–Li provides a sharp contrast, and exemplifies the modern trend towards the use of these reagents.

4 Alkanes

Organic nomenclature, the naming of compounds, is based on the alkanes' names. This is discussed in Section 4.1. The chemistry of the alkanes forms Section 4.2.

4.1 Systematic nomenclature

Many compounds have attractive traditional (trivial) names which are short and handy for everyday use. For publications, however, names denoting compounds' structures unambiguously are essential. These systematic names are constructed by a rigid procedure the full form of which, replete with its arcane conventions, is exceedingly difficult. Unfortunately, a truly 'simple' account does not bring out the underlying principles. The present version is a compromise. *It covers open–chain (aliphatic) compounds fairly comprehensively but cyclic compounds, apart from the cycloalkanes, are not discussed.* Even with open–chain systems the going is hard enough but, some comfort, the details need not be memorised.

Scheme **4.1Ge** deals first with alkanes (**part 1**), then with compounds (**part 2**) containing O, N, Hal, S,.....A range of examples follows.

4.1Ge

Skeleton formulae are used to save space e.g.:

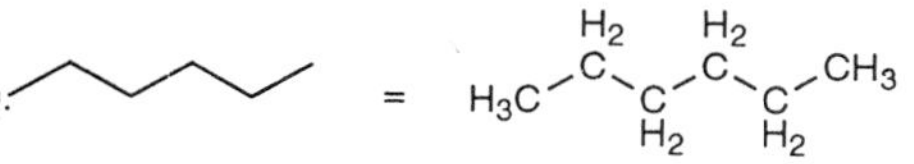

$H_3C - \underset{H_2}{C} - \underset{}{C} - \underset{H_2}{C} - CH_3$

1 Alkanes

Aliphatic C_nH_{2n+2}

n = 1	2	3	4	5	6	7	8	9	10
methane	ethane	propane	butane	pentane	hexane	heptane	octane	nonane	decane etc

Cyclic $(CH_2)_n$

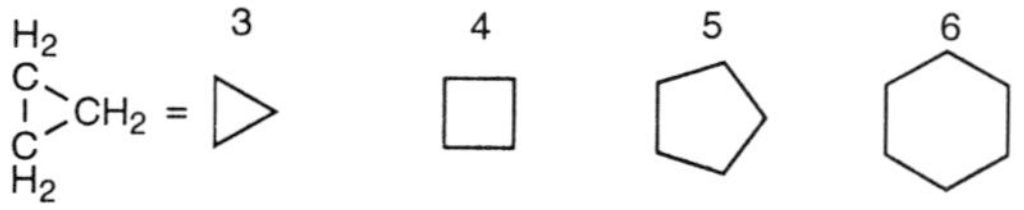

cyclopropane cyclobutane cyclopentane cyclohexane etc

Alkyl groups, change —ane to —yl
e.g. propyl $CH_3–CH_2–CH_2–$

Alkanes with branched chains:

Number the longest chain in the direction that gives the lowest possible numbers to the side chains. This gives the basic name. Introduce the side chains as prefixes (i.e. before the basic name) in alphabetical, not numerical, order.

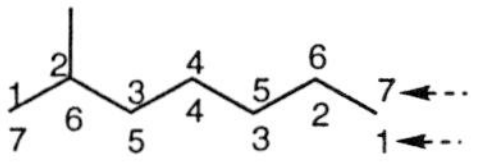

correct
incorrect

2–methylheptane

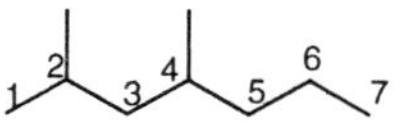

2,4–dimethylheptane (di = two)

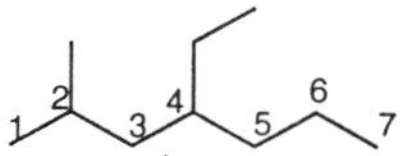

4–ethyl–2–methylheptane

#(for future reference)

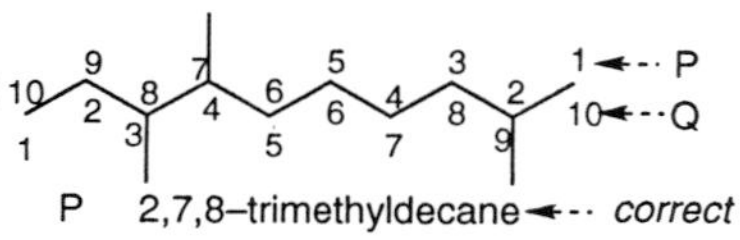

P 2,7,8–trimethyldecane *correct*
Q 3,4,9–trimethyldecane *incorrect*

Q gives a lower total (16) than P (17). There is a widespread misconception that the name with the lower total (here Q) is correct. There is a more important rule, that the name having the lower number at the 'point of first difference' is preferred. Here the first difference is 2 in P versus 3 in Q. Hence P is correct.

2 Compounds containing O, N, Hal, S.......(i.e. most organic compounds)

Some functional groups are named only as prefixes, some as prefixes or suffixes (i.e. at the end of the name), and two only as suffixes. (#With cyclic compounds the situation is more complicated.) A name is allowed to have only one suffix. Groups containing O, N, Hal, S, etc are arranged in a priority order for selection as the *Principal Function*. The following Table shows the order, and deals with a tricky point about the endings –ene and –yne

To devise a name: **a** select the Principal Function, **b** use this as the suffix, **c** give it (or the C to which it is attached) the lowest possible number and so deduce the basic name, **d** specify the other groups as numbered prefixes

Table of groups,(i) to (xv), in descending order of priority for selection as **Principal Function**

	Group	Suffix	Prefix		Group	Suffix	
(i)	acid –CO_2H	–oic acid					These replace –ane of basic name
(ii)	anhydride –CO–O–CO–	–oic anhydride		(xi)	C=C	–ene	and are regarded as part of basic name. Do not 'count' as suffixes.
(iii)	ester CO_2R	R –oate / name	R–oxycarbonyl	(xii)	–C≡C–	–yne	Are followed by suffix of any higher priority group in compound
	see e.g.s for clarification						
(iv)	acyl halide –CO–Hal	–oyl halide	halocarbonyl–				
(v)	amide –CO–N	–amide	amido–				
(vi)	nitrile –CN	–nitrile	cyano–				Prefix only
(vii)	aldehyde –CO–H	–al		(xiii)	ether –OR		R–oxy
(viii)	ketone –CO–	–one	oxo– (denoting =O)				(see e.g.s)
(ix)	alcohol –OH	–ol	hydroxy–	(xiv)	halides –Hal		halo–
	amine –N	–amine	amino–	(xv)	nitro –NO_2		nitro–

The following e.g.s show how the procedures are applied. Do not try to memorize the details.

a | **butanoic acid**

Idea is one CH_3 of butane is converted, in theory, into CO_2H. A CO_2H must be at end of chain, and its C is always numbered 1. Thus no need to show 1 in name. Name should be butaneoic but, convention, e is dropped if vowel or y follows.

b | **3–hydroxybutanoic acid**

The Principal Function (PF) is CO_2H, hence suffix is –oic acid. The OH is named as a prefix, and its number must be specified.

c | **butanedioic acid**

Di before acid, 2 acid groups.Both CO_2H must be at end of chain, so no need for numbers. Traditional name *succinic acid* still in vogue.

d | **propanoic anhydride**

Product of removing H_2O from 2 molecules of propanoic acid. Word anhydride means loss of H_2O from some compound.

e | **ethyl butanoate**

separate words

Full structure

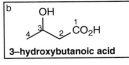

Ethyl ester of butanoic acid

f | **ethyl 3–oxobutanoate**

Is a ketone and an ester. Ester is PF. Number the C chain. Regard keto group as a substituent O=(C) and specify as a prefix. Traditional name *acetoacetic ester* generally used.

g | **diethyl propanedioate**

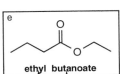

Full structure

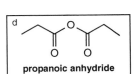

Ester of propanedioic acid. Two ester groups so must specify both ethyl. Traditional name *malonic ester* still generally used.

h | **N,N–dimethylmethanamide**

Italicised *N,N* show that both Me are attached to N.Traditional name *dimethylformamide* and abbreviation *DMF* generally used.

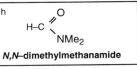

i | **ethanamide**
j | **ethanenitrile**

Traditional names *acetamide*(i) and *acetonitrile*(j) stubbornly defy extinction.

k

hex–5–en–2–ol

–ene regarded as part of basic name, not a suffix. OH, the PF, used as suffix does not contain C. So C to which OH attached given lowest possible number. Double bond is 5,6. Lower number cited. Although there is a number between ene and ol final e of ene is still omitted.

l

hex–4–yn–2–one

Keto group given lowest possible number. The e of –yne is omitted.

m

1–ethoxybutane

Shorter RO regarded as substituent, so not butyloxyethane. **RO groups:**

MeO methoxy ⎫ The yl ending of R
EtO ethoxy ⎭ name is dropped

PrO propyloxy ⎫ yl is retained
BuO butyloxy etc ⎭

n

7–bromo–5–nitroheptan–3–amine

NH_2 is PF, not contain C. So C to which attached given lowest possible number. Would be 5 if numbered from other end. Prefixes cited alphabetically.

4.2 Chemistry

Alkanes do not react with electrophiles, nucleophiles, acids or bases unless very vigorous conditions or very reactive reagents are used; they do react with radical reagents (**4.2Ge**).

4.2Ge

The preparations of alkanes are in **4.2Pr**

Most of the material in **4.2Pr** does not refer to preparations in the usual sense. These days no chemist would set about preparing, say, decane. Standard alkanes are supplied, cheaply and in a pure condition, by the petroleum industry. For various purposes, however, (for example in natural work) it is necessary to remove the functional groups of compounds: methods for doing so are in **Pr1**. Synthetic work sometimes involves the conversion of a compound R–(functional group) into R–R or R–R¹: this is covered in **Pr2**.

The important process of hydrogenation(**Pr1a**) is usually carried out using heterogeneous catalysis, i.e. reaction on the insoluble metal surface. Homogeneous catalysts are now available and have the advantage of greater selectivity. Thus, when double bonds of different types are present a particular one can be reduced leaving the others intact.

Removal of a CO double bond is shown in **Pr1b.** Procedures of type (**i**), and the sequence (**ii**) involving the thioacetal are convenient and reliable. Less useful methods, for example the Clemmensen reaction, have been omitted. The best method for converting R–Hal into R–R or R–R¹ (**Pr2**) involves the lithium cuprates (**3Pr2a**). Two old preparations of R–R (the Wurtz reaction using R–Hal and the Kolbe reaction using R–CO₂H) have been superseded and can be disregarded.

4.2Pr

1 Reduction

a Hydrogenation of alkenes and alkynes

$$\ce{C=C} \xrightarrow[\text{H}_2 \text{ (possibly at high pressure) / 20–150°}]{\text{solvent (e.g. EtOH) / catalyst / stir under}} -\overset{|}{\underset{H}{C}}-\overset{|}{\underset{H}{C}}-$$

$$-C\equiv C- \xrightarrow{\hspace{5cm}} -\overset{|}{\underset{H_2}{C}}-\overset{|}{\underset{H_2}{C}}-$$

heterogeneous catalysts: Pd–C (Pd adsorbed on charcoal); Pt (in a finely divided state);

Raney Ni (prepared cheaply in a finely divided state by treating

Ni / Al alloy with NaOH / H₂O)

less active, 'poisoned' $\begin{cases} \text{Pd–CaCO}_3\text{–Pb(QAc)}_2 \text{ Lindlar catalyst} \\ \text{Pd–BaSO}_4 \text{ –quinoline} \end{cases}$ poisons

homogeneous catalysts: e.g. (Ph₃P)₃RhCl

Simple representation (#more in later courses):

e.g. H₂ and CH₂=CH₂ are adsorbed on a metal surface ⟶

H–M and C–M bonds formed

one H transferred to C,
CH₃ has little affinity for M

CH₃–CH₃ is ⟵ [diagram] ⟵ second H ⟵ [diagram]
desorbed from M   transferred

The two H atoms are added to the same side of the double bond (see **6Re1**)

b Aldehydes and ketones

$$\ce{C=O} \longrightarrow \ce{CH_2}$$

nothing special about this;
just an alcohol of high b.p.

(i) H₂N–NH₂ (hydrazine) / KOH / HO–(CH₂)–O–(CH₂)–OH / Δ

a convenient version of the original Wolff–Kishner reaction (mechanism in margin)

Widely used modern developments include:

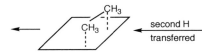

$$\ce{C=O} \xrightarrow{\text{H}_2\text{N–NH–Ts}} \ce{C=N–NH–Ts} \xrightarrow{\text{NaBH}_4\text{ / EtOH}} \ce{CH_2}$$
7.4Re2c(ii)

(ii)

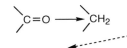

$$\ce{C=O} \xrightarrow[\text{sq BF}_3]{\text{HS–(CH}_2)_2\text{–SH /}} \underset{\text{(ethane–1,2–dithiol)}}{[\text{dithiolane}]} \xrightarrow[\text{(has H}_2\text{ on surface from Pr)}]{\text{Raney Ni / }\Delta} \ce{CH_2}$$
7.4Re2b(iii)

c R–Hal and R–OH ⟶ R–H **2.1Re4**

see **8.2Re2** for R–CO₂H ⟶ R–H

2 Alkyl halides ⟶ R–R or R–R¹ **3 Pr 2a(ii)**

LiCuR₂ + R¹–Br $\xrightarrow{0°}$ R–R¹

(R and R¹ may be identical or different)

$$R-\overset{-}{C}u-R \quad R^1-Br \quad S_N2$$

---

mechanism:**Pr1b(i)**

$$\ce{C=O} \longrightarrow \underset{\textbf{7.4Re2c(ii)}}{\ce{C=N–NH_2}}$$

↓ HO⁻

$$\ce{CH–N=N–H} \quad HO^-$$

↓

$$\ce{CH–N=N^-}$$

H–OR (the high b.p. alcohol)

↓

$$\ce{CH_2} + N_2 + RO^-$$

modern development

$$\ce{C=N–NH–Ts}$$
H–$\overset{-}{\text{B}}$H₃

↓

$$\ce{CH–N=NH}$$

↓ as above

$$\ce{CH_2}$$

The alkanes' reactions are covered in **4.2Re**.

4.2Re

1 With Hal$_2$ (reminder, that Hal$_2$ represents F$_2$, Cl$_2$, Br$_2$, I$_2$)

a methane

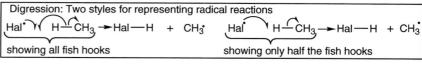

Digression: Two styles for representing radical reactions

Hal$^{\bullet}$ H$-$CH$_3$ →Hal$-$H + CH$_3^{\bullet}$ Hal$^{\bullet}$ H$-$CH$_3$→Hal$-$H + CH$_3^{\bullet}$

showing all fish hooks   showing only half the fish hooks

CH$_4$ +   Hal$_2$ $\xrightarrow{\Delta \text{ or } h\upsilon}$ CH$_3$–Hal +   H–Hal   (hυ means irradiate with UV light)

Hal$-$Hal $\underset{\Delta \text{ or } h\upsilon}{\rightleftharpoons}$ 2 Hal   initiation

Hal$^{\bullet}$ + H$-$CH$_3$→Hal$-$H + CH$_3^{\bullet}$
                                                        } propagation
CH$_3^{\bullet}$ + Hal$-$Hal →CH$_3$–Hal + Hal$^{\bullet}$

radical chain reaction
need sq of Hal$^{\bullet}$ to start
reaction, Hal$^{\bullet}$(chain carrier)
regenerated in propagation
steps which are repeated
until all CH$_4$ used

Calculate enthalpy change from bond energies

Table of **Bond Energies** (kJ mol^{-1})

F–F	157	H–F	568	CH$_3$–F	456	H–H	435
Cl–Cl	243	H–Cl	431	CH$_3$–Cl	351	H–OH	502
Br–Br	194	H–Br	365	CH$_3$–Br	293	H–NH$_2$	427
I–I	153	H–I	299	CH$_3$–I	234		

C—H bonds:

R$_3$C—H	381	R$_2$CH—H 397		RCH$_2$—H 410	CH$_3$—H 435
CH$_2$=C–C–H 372		Ph–CH$_2$–H 355		Ph–H 460	

Thus:

CH$_4$ + Cl$_2$ → CH$_3$Cl + HCl

breaking CH$_3$–H breaking Cl–Cl

ΔH_o = −431 + 435 − 351 + 243 = −104 kJ

forming H–Cl forming CH$_3$–Cl

Similarly for all Hal:

Hal	F	Cl	Br	I
ΔH_o	−432	−104	−29	55

} F too reactive, Cl and Br useful, I unreactive

b higher alkanes e.g. % attack at sites of 2–methylbutane (1 mol) by Cl$_2$ (1 mol) at 250°

CH$_3$ ◄--16

16 --►H$_3$C—C—C—CH$_3$◄- 16
         | / \
         H H H
         ▲   ▲
         ¦   ¦
        23 29

	CH$_3$	CH$_2$	CH
number of H atoms	9	2	1
% attack	48	29	23
% attack at one H	5.3	14.5	23
relative reactivity of one H	1.0	2.7	4.3

c Br$_2$ versus Cl$_2$

Generalisation: *As reactivity of system increases selectivity of reaction decreases*
Br$_2$ is less reactive, hence more selective: in relation to C–H bond breaking, the transition state is earlier for Cl$_2$ than for Br$_2$

2 Vapour phase nitration

$$\text{R—H} \xrightarrow{\text{HNO}_3 / 450°} \text{R—NO}_2 + \text{R}^1\text{—NO}_2 + \text{R}^2\text{— NO}_2 \quad \text{etc}$$

mixture formed, in some products R has been degraded (R^1 and R^2 have fewer C atoms than R); one of main uses is to produce CH$_3$–NO$_2$ nitromethane (**11Re1**)

Two representations of a step in **Re1a** are shown. Which to use is a matter of personal choice; the chemistry is not affected. The enthalpy change of halogenation may be calculated from the bond energies given in the table. (The value for a specific bond e.g. CH$_3$–Cl is, strictly, a bond–dissociation energy. Bond energy refers to the average value in a series of similar compounds e.g. the C–Hal values in **2.1Ge2b**. The present table contains both types, but we can ignore the distinction.) Calculation of the enthalpy changes with the different Hal$_2$ explains the observations that I$_2$ does not react with alkanes whereas F$_2$ reacts violently and generally in an uncontrollable manner. Chlorination of 2–methylbutane(**1b**) gives much more 2–chloro–2–methylbutane than expected on a random basis. This illustrates the influence of radical stability in determining the proportions of products: the relative stabilities of C radicals (Section 1.9), tertiary > secondary > primary, correspond with the C–H bond energies of the systems from which they are formed. Unfortunately results are not available for chlorination and bromination of 2–methylbutane under the same conditions. Bromination at 25°(**1c**) is far more selective than chlorination at 250°. No doubt the lower temperature accounts for much of this difference, but another factor is the lower reactivity of Br$_2$. A generalisation relating selectivity to reactivity is given below **1c**; from this it follows, for example, that reaction **1b** at very high temperature would give the purely statistical result of 75% attack at CH$_3$, 16.7% at CH$_2$, 8.3% at CH. ($^\#$Mechanism course for discussion of earlier transition state.) The conditions of nitration in **Re2** are very different from those of the familiar nitration of aromatic compounds.

5 Alkenes

Before dealing with the alkenes we should consider the general relations between alkanes, alkenes and alkynes (scheme **5Ge1**).

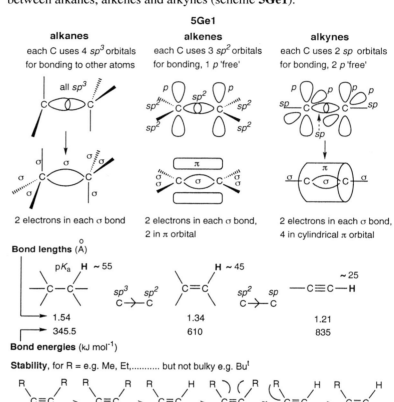

5Ge1

alkanes	alkenes	alkynes
each C uses 4 sp^3 orbitals for bonding to other atoms	each C uses 3 sp^2 orbitals for bonding, 1 p 'free'	each C uses 2 sp orbitals for bonding, 2 p 'free'

| 2 electrons in each σ bond | 2 electrons in each σ bond, 2 in π orbital | 2 electrons in each σ bond, 4 in cylindrical π orbital |

Bond lengths (Å)

pK_a H ~ 55   H ~ 45   ~ 25

sp^3 sp^2 sp^2 sp
$C \rightarrow C$ $C \rightarrow C$

1.54   1.34   1.21
345.5   610   835

Bond energies (kJ mol^{-1})

Stability, for R = e.g. Me, Et,........... but not bulky e.g. Bu^t

In the very simple molecular orbital pictures only the CC bonds are shown in detail. ('Foundations' gives a clear description of the background.) The π electrons are not involved in the molecular framework; they are relatively distant from the C atoms and available for reaction. Thus the first general characteristic of alkenes and alkynes is their *tendency to be attacked by electrophiles.*

The spherical *s* orbital of C may be regarded as near to the positive nucleus, and the pear shaped *p* orbital as farther away. (#*s* has a high probability value at the nucleus whereas *p* has a node.) It follows that as the *s* character of an orbital increases the orbital becomes effectively nearer the nucleus. An electron in an orbital therefore becomes more strongly attracted by the positive nucleus as the orbital goes from sp^3 to sp^2 to sp. If 2 electrons are in a C–H bond departure of a proton leaves a carbanion which becomes more stable in the same hybridisation order, i.e. the acidity of the H increases. If the electrons are in a bond to another C they are attracted in the directions shown in **5Ge1**: this is the basis of the inductive effects of differently hybridised C (**1.6Ge**). The inductive effect is concerned only with σ electrons: the presence of an excess of electrons in the π orbitals of CC multiple bonds is irrelevant. Electron donation by an R group attached to multiple bond is a mutually satisfying stabilising effect. A double bond therefore becomes more stable as more R groups are attached to it (bottom of scheme). However in 1,1–di–, cis–1,2–di–, tri– and tetra–alkyl systems there is the potential for repulsion between the R groups. This is small for Me–Me interactions (~4 kJmol^{-1}) but with Bu^t groups (**2.1Ge2**, margin) steric factors are dominant and the stability order in the scheme does not apply. For example the tetra Bu^t alkene is so strained that its preparation is very difficult.

Relative bond lengths and energies are not readily predicted by simple organic theory. From the energies it follows that additions to multiple bonds are energetically very favourable (material in margin). Thus the second characteristic of alkenes and alkynes is their tendency *to undergo addition reactions*.

Scheme **5Ge2** is concerned with a stereoelectronic effect, and the direction of eliminations.

ΔH°(kJ mol^{-1}) of reactions

$$H_3C - CH_3 \ + \ Br_2$$
$$\downarrow \qquad -29$$
$$2CH_3Br$$
not a real reaction

$$H_2C = CH_2 \ + \ Br_2$$
$$\downarrow \qquad -113$$
$$BrH_2C - CH_2Br$$

$$HC \equiv CH \ + \ Br_2$$
$$\downarrow \qquad -150$$
$$BrHC = CHBr$$

5Ge2

A stereoelectronic effect

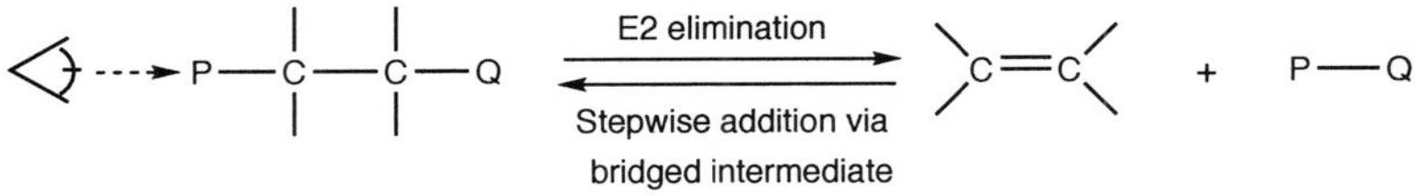

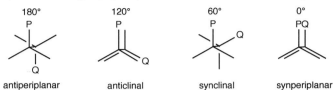

Conformations, and dihedral angles (angles eye sees between C—P and C—Q bonds)

180°	120°	60°	0°

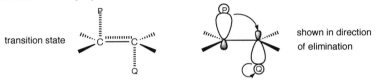

antiperiplanar   anticlinal   synclinal   synperiplanar

The effect   Antiperiplanar conformation favoured for reaction in both directions

transition state   shown in direction
of elimination

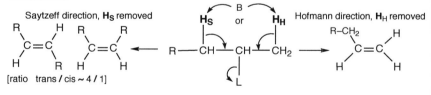

Direction of E2 and E1 eliminations

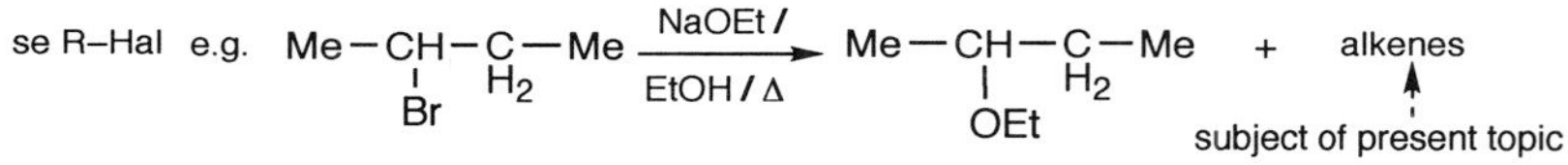

se R–Hal   e.g.   Me—CH—C—Me $\xrightarrow[\text{EtOH}/\Delta]{\text{NaOEt}/}$ Me—CH—C—Me   +   alkenes
                     | H_2 | H_2
                   Br   OEt   subject of present topic

E2 reactions

Saytzeff direction, H_S removed   B   Hofmann direction, H_H removed

[ratio   trans / cis ~ 4 / 1]

for R = Me, B = NaOEt / EtOH,

L =	Br	F	N^+Me_3
% Hofmann	19	80	95
% Saytzeff	81	20	5

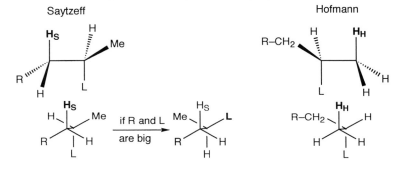

Hofmann / Saytzeff ratio increases as
  (i) strength of B increases
  (ii) as electron withdrawal by L increases

  (iii) as L$^-$ becomes a poorer leaving group
  (iv) as size of B increases
  (v) as size of R or L increases

Conformations with H (to be eliminated) and L in antiperiplanar relation

E1 reactions

high yield of 2–methylbut–2–ene irrespective of nature of R

If conditions are changed so that **E2** is favoured

| RO$^-$ = MeO$^-$ | 81% | 19% |
| RO$^-$ = Et$_3$O$^-$ | 22% | 78% |

Possible conformations of a disubstituted ethane are shown. [It would be logical to use the terms *syn* (same side) and *anti* (opposite sides) only for reactions, and the corresponding terms *cis* and *trans* for compounds. Unfortunately syn and anti have been incorporated into the conformations' names and distinction between the two sets has become blurred.] The important point is that *the antiperiplanar conformation is favoured for the substrate in E2 and the product in the additions specified.* Two factors are involved. An antiperiplanar arrangement of the atomic centres allows maximum overlap of the developing orbitals in the formation of, or in the addition to, the π bond. The anti orientation of P and Q also minimises steric interaction in the elimination or the addition. Caution: this generalisation does not apply to E2 in cyclic systems of 4, 5 (possibly), 8, 9 and 10 members where synperiplanar elimination is favoured.

Only a brief treatment of the E2 direction is given here. ($^{\#}$Full discussion, later courses.) In general secondary RHal give comparable amounts of substitution and elimination products (**2.1Ge2c**). Here only the elimination is considered but the total yield of alkenes may be 50% or less. An excess of base is present, and the elimination may taken as E2 even if this is not strictly so. (See the discussion of **2.1Ge2**.)

Dehydrobromination of, for example, 2–bromobutane may proceed in 2 directions giving but–1–ene or but–2–ene. Several features determine which is formed. Saytzeff (1841–1910) studied dehydrohalogenation while Hofmann (1818–1894, a giant of classical organic chemistry) investigated elimination from quaternary ammonium salts: their names are used to designate the directions of elimination. In the general formula (scheme) H_H and H_S are, respectively, the H removed in the Hofmann and Saytzeff directions. H_H is more acidic than H_S (Section 1.9). H_H should then be removed and the Hofmann direction, giving the less substituted alkene, should be followed. However the Saytzeff product is the more substituted and therefore the more stable alkene. In a simple system (leaving group L=Br, small base B, small R) the second factor is dominant and the Saytzeff direction may be regarded as the 'natural' one. Various factors may prompt a change towards Hofmann (scheme). *(i)* As the strength of base increases proton removal becomes the dominant factor and Hofmann is relatively more favoured. *(ii)* Strong electronic withdrawal by L makes both H_H and H_S more acidic. Proton removal again becomes the dominant factor. However H_S is less affected than H_H because enhanced +I donation by R can reduce the effect on H_S (sketch). The upshot is that Hofmann becomes more favoured. [$^{\#}$In (i) and (ii) the elimination has some E1cb character.] *(iii)* As L$^-$ becomes a better leaving group its departure may be regarded as marginally ahead of the base attack. From the treatment of E1 reactions (following paragraph) it emerges that this change favours the Saytzeff direction. ($^{\#}$The elimination has some E1 character.) So then the reverse change, making L$^-$ a poorer leaving group, favours Hofmann. For example, F is strongly electron withdrawing and a poor L$^-$; both properties contribute to the remarkable difference in product ratio between F and Br. *(iv)* On steric grounds the base prefers H_H; attack on H_S brings the base into proximity with the R of the αC. As the base becomes bigger the preference is enhanced, hence more Hofmann. *(v)* The antiperiplanar conformation required for Saytzeff has R and L in a synclinal relation (sketch). If either R or L is big, or if both are big, the repulsion between these groups drives the molecule into other conformations in which H and L are no longer antiplanar. The preferred Hofmann conformation is not affected by the size of R and L, and the Hofmann percentage increases.

In elimination from tertiary substrates high yields of alkenes are obtained (**2.1Ge2a**). Under standard conditions, protic polar solvents, E1 occurs. Loss of H from a carbanion (scheme) is a fast step (very low activation energy) and is therefore governed by product stability. Thus E1 leads mainly or exclusively to Saytzeff. However, nonpolar aprotic solvents discourage formation of ions and even tertiary substrates can be forced towards E2. Change from Saytzeff towards Hofmann in E2 is again influenced to some degree by structural features but only an increase in the size of base has a marked effect. (Scheme, and a preparatively useful example **2.1Re2**.)

The main routes to alkenes are in scheme **5Pr**.

5Pr

1 Aldehydes and ketones

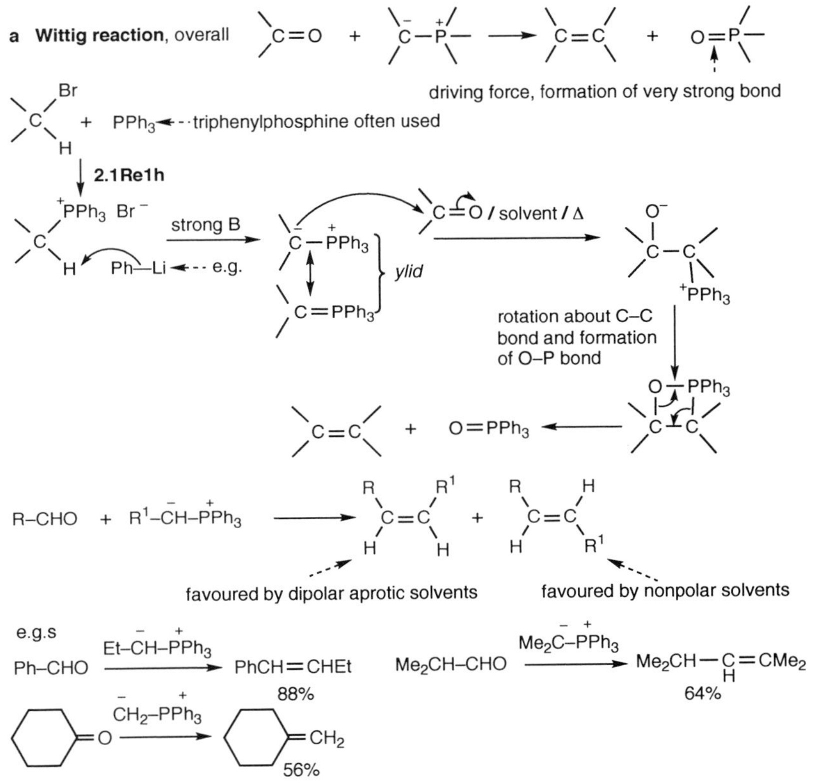

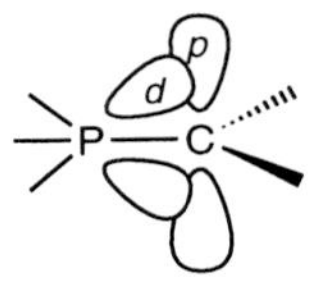

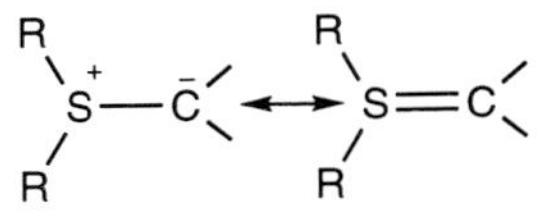

a Wittig reaction, overall

driving force, formation of very strong bond

PPh$_3$ ◀‥triphenylphosphine often used

2.1Re1h

strong B

Ph—Li ◀‥ e.g.

ylid

C═O / solvent / Δ

rotation about C–C bond and formation of O–P bond

$$C{=}C \quad + \quad O{=}PPh_3 \longleftarrow$$

$$R{-}CHO \quad + \quad R^1{-}\overset{-}{C}H{-}\overset{+}{P}Ph_3 \longrightarrow$$

favoured by dipolar aprotic solvents   favoured by nonpolar solvents

e.g.s

$$Ph{-}CHO \xrightarrow{\ Et{-}\overset{-}{C}H{-}\overset{+}{P}Ph_3\ } PhCH{=}CHEt$$

88%

$$Me_2CH{-}CHO \xrightarrow{\ Me_2\overset{-}{C}{-}\overset{+}{P}Ph_3\ } Me_2CH{-}\underset{H}{C}{=}CMe_2$$

64%

$$\text{cyclohexanone} \xrightarrow{\ \overset{-}{C}H_2{-}\overset{+}{P}Ph_3\ } \text{methylenecyclohexane}$$

56%

[in general formulae of the following type R groups may be identical or different]

yields of	RCH=CH$_2$ and RCH=CHR	R$_2$C=CH$_2$ and RCH=CR$_2$	R$_2$C=CR$_2$
	high	medium (~50–65%)	low (steric hindrance)

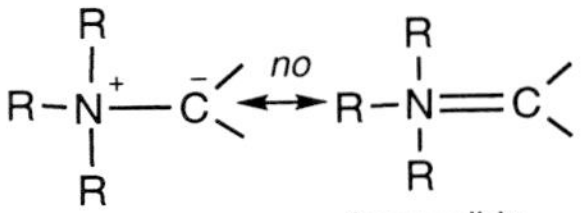

b McMurray reaction

$$C{=}O \; + \; O{=}C \xrightarrow{\ Ti^{\circ} / \Delta\ } C{=}C \; + \; TiO_2$$

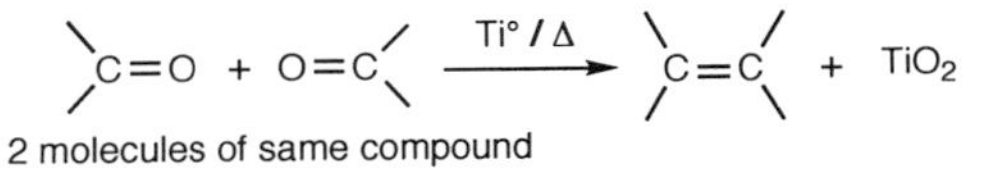

Ti° represents colloidal Ti, generated from TiCl$_4$ / LiAlH$_4$

2 molecules of same compound

$$\begin{array}{ccc} R & & R \\ \diagdown\!\!\overset{+}{S}{-}\overset{-}{C}\diagup & \longleftrightarrow & \diagdown S{=}C\diagup \\ \diagup & & \diagup \\ R & & R \end{array}$$

$$\begin{array}{ccc} R & & R \\ | & no & | \\ R{-}\overset{+}{N}{-}\overset{-}{C}\diagdown & \longleftrightarrow & R{-}N{=}C\diagdown \\ | & & | \\ R & & R \end{array}$$

impossible, 10 electrons in L shell

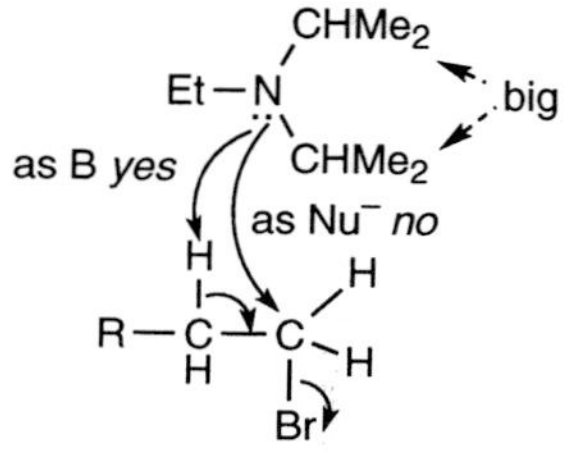

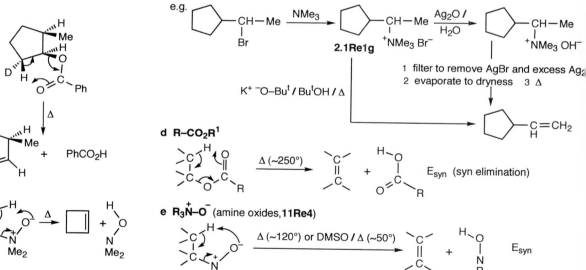

2 1,2–Elimination from $H-\overset{|}{\underset{|}{C}}-\overset{|}{\underset{|}{C}}-L$

a R–Hal or R–OTs / B   typical conditions Na⁺ (or K⁺) ⁻OR / ROH or PhMe / Δ

see **2.1Re2** for an example

for pm R–Hal need special reagents e.g. Hünigs base   ---→ Et–N(CHMe₂)₂

$$Me-(CH_2)_5-CH_2-CH_2-Br \xrightarrow{\text{Hünigs base} / \Delta} Me-(CH_2)_5-CH=CH_2$$

yield 95%; remarkable!

b R–OH / A / Δ   A = e.g. H_2SO_4, BF_3, $ZnCl_2$ tendency for elimination te > se > pm

E1 or E2 mechanism, usually Saytzeff direction, more substituted alkene formed

Polyphosphoric acid (H_3PO_4 containing P_2O_5) often better than H_2SO_4

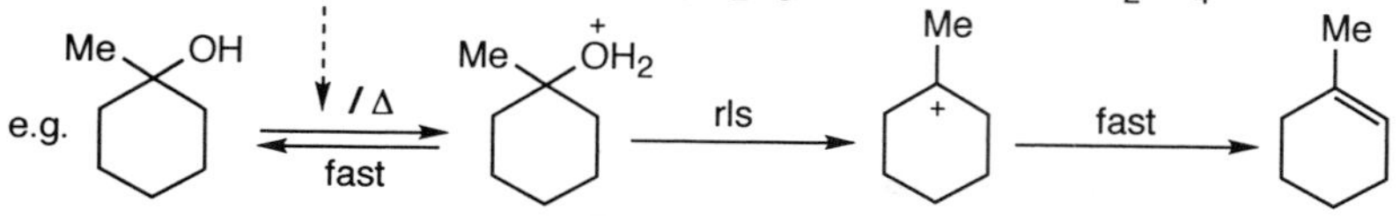

e.g.

E1, regioselective

For se ROH better to use method **a** rather than method **b**

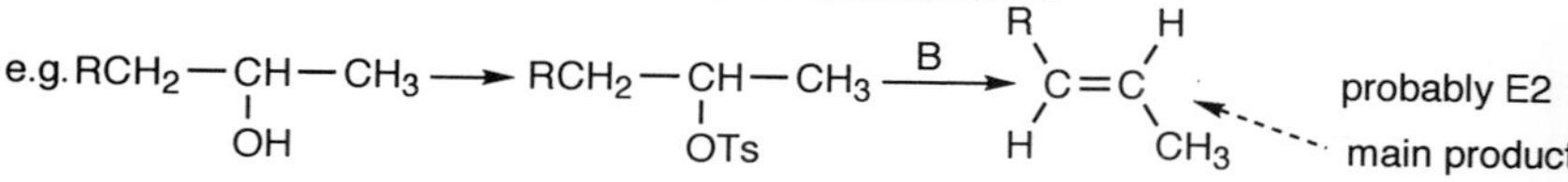

e.g. $RCH_2-CH(OH)-CH_3 \longrightarrow RCH_2-CH(OTs)-CH_3 \xrightarrow{B}$ C=C   probably E2 ···· main product

c R_4N^+ OH⁻ / Δ or R_4N^+ X⁻ / B / Δ

e.g.

$\longrightarrow$ CH–Me with Br $\xrightarrow{NMe_3}$ CH–Me with $^+NMe_3$ Br⁻ $\xrightarrow{Ag_2O / H_2O}$ CH–Me with $^+NMe_3$ OH⁻

2.1Re1g

K^+ ⁻O–Buᵗ / BuᵗOH / Δ

1 filter to remove AgBr and excess Ag₂

2 evaporate to dryness   3 Δ

$\longrightarrow$ C=CH₂

d R–CO₂R¹

$\xrightarrow{\Delta (\sim 250°)}$ C=C + $\overset{H-O}{\underset{R}{\overset{|}{C}=O}}$ E_syn (syn elimination)

e $R_3\overset{+}{N}-O^-$ (amine oxides, 11Re4)

$\xrightarrow{\Delta (\sim 120°) \text{ or DMSO} / \Delta (\sim 50°)}$ C=C + $\overset{H-O}{\underset{R_2}{N}}$ E_syn

Cope elimination

3 1,2– Elimination from $L-\overset{|}{\underset{|}{C}}-\overset{|}{\underset{|}{C}}-L^1$

a 1,2– dibromides

b anions of 2–bromoacids

4 $R-C\equiv C-R^1$ 4.2Pr1a

The Wittig reaction (**5Pr1a**,1953) is one of the most important reactions of organic chemistry. It joins compounds together simply and cleanly, giving alkenes which serve as starting materials for further transformations shown in **5Re**. 1,2–Disubstituted alkenes are generally formed as *E /Z* mixtures with *Z* predominating. [The terms *E* and *Z*, which denote the configurations of compounds containing a double bond, will be explained in a stereochemistry course. For symmetrical alkenes *E* and *Z* are equivalent to the familiar terms trans and cis respectively.] Several factors influence the stereochemical outcome, as exemplified in the scheme by the effect of changing solvent polarity. ([#]Full treatment,later courses.) The reactive intermediate (scheme) is an ylid, pronounced ill–id not eye–lid. One canonical (Section 1.6) of an ylid has C^- joined to an atom such as P^+, S^+, or N^+. Resonance stabilisation (arising from a second, neutral canonical) is possible with P and S but not with N, and N ylids are much less stable.

Pr2 and **3** are eliminations. That a primary R–Br undergoes elimination rather than substitution with Hünigs base (**2a**) runs counter to the generalisation of **2.1Ge2**. However this base is big, apparently too big to act as a nucleophile but still able to act as a base. (See diagram in margin, and rationalisation in **1.12Ge**.) The main features of **2b** and **2c** accord with the treatment of **5Ge2**. **Pr2d** and **2e** represent a class of elimination in which the substrate is heated strongly without reagent or solvent in an inert atmosphere (N_2) or at reduced pressure. These are examples of *pyrolysis*, from *pyro* = heat and *lysis* = splitting, and are designated E_{syn}.They could with propriety be included in the E1 class, but they are so different in nature from the usual E1 reactions that it is sensible to give them a separate title. Study of isotopically labelled substrates, for example the ester in the margin, prove the syn stereochemistry of **2d**. **2e** occurs at a lower temperature than **2d** and is useful for preparing strained alkenes such as cyclobutene.

In **Pr3** reaction of the substrates in the antiperiplanar conformations (scheme) minimises the energies of activation. The reactions are *stereospecific*. (A stereospecific reaction is one in which stereoisomers of a substrate give, selectively, stereoisomers of a product. Replacement of 'stereoisomers' by 'diastereoisomers' is strictly correct but pedantic.) It is instructive to draw sketches for the material in the margin opposite **3a**; points of stereochemistry and nomenclature are involved. This reaction is usually employed in the reverse direction, i.e. the dibromides are obtained from the alkenes, as shown in the following scheme (**5Re**).

Alkenes exhibit a wide range of synthetically useful reactions. The scheme showing these (**5Re**) is necessarily long. Don't try to work through it in one sitting. Take it slowly, and before starting look at the commentary which follows the scheme.

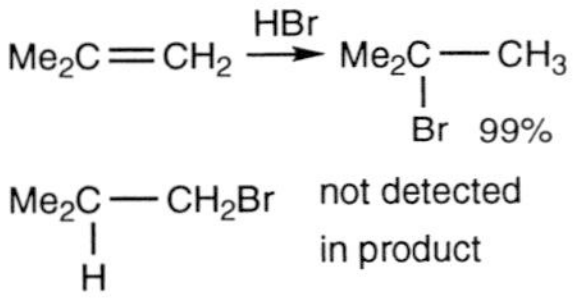

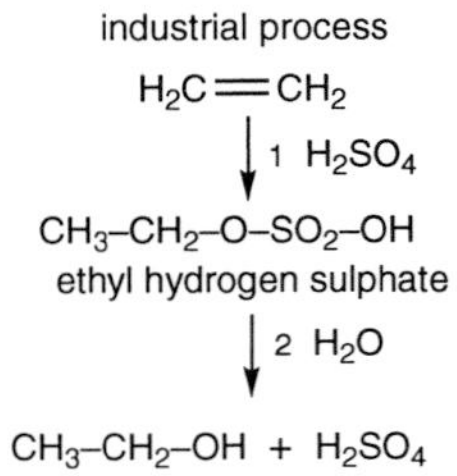

5Re

Addition of E$^+$ to equivalent faces of simple alkene

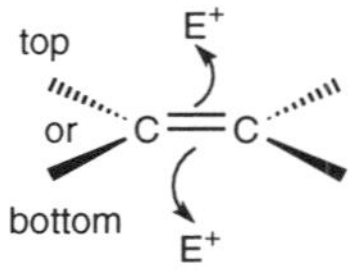

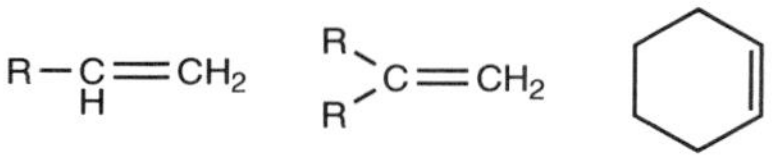

used for illustrating various reactions

Me$_2$C=CH$_2$ $\xrightarrow{\text{HBr}}$ Me$_2$C—CH$_3$
                                             |
                                             Br 99%

Me$_2$C—CH$_2$Br   not detected
        | in product
        H

In an equilibrium:

Me$_2$C—CH$_3$ 81%
        |
        Br

Me$_2$C—CH$_2$Br   19%
        |
        H

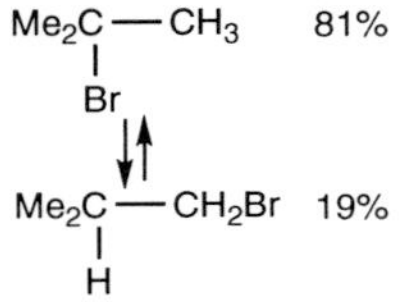

Thus HBr addition is under
kinetic control (Section **1.5**)

Re1 and **2** are represented

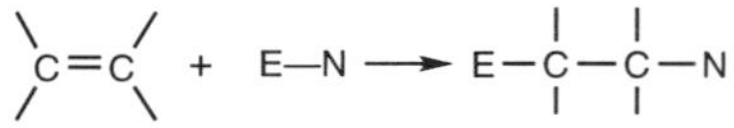

$\diagdown$C=C$\diagup$ + E—N ⟶ E—C—C—N

With unsymmetric alkenes **Re1**, **Re2** and **Re4a** are regiospecific

1 Stepwise E$^+$ addition, intermediate not bridged

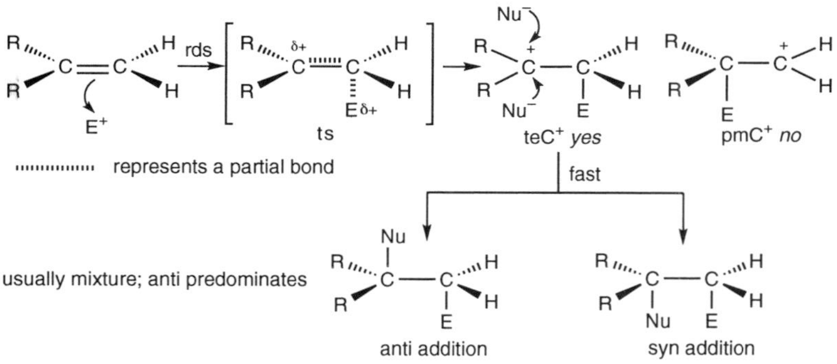

‧‧‧‧‧‧‧‧‧‧‧ represents a partial bond

usually mixture; anti predominates

anti addition   syn addition

industrial process

H$_2$C=CH$_2$
    | 1 H$_2$SO$_4$
    ↓
CH$_3$–CH$_2$–O–SO$_2$–OH
ethyl hydrogen sulphate
    | 2 H$_2$O
    ↓
CH$_3$–CH$_2$–OH + H$_2$SO$_4$

a H–Hal

reactivity HI > HBr > HCl > HF ◄‧‧‧ use HF /
              slightly slightly

e.g. R—C=CH$_2$ $\xrightarrow[\text{rds}]{\text{HBr / THF / 20°}}$ R—C—CH$_3$ $\xrightarrow{\text{fast}}$ R—CH—CH$_3$
        | | |
        H   H   Br
          ↖H$^+$ (THF = tetrahydrofuran, Section **1.13**)

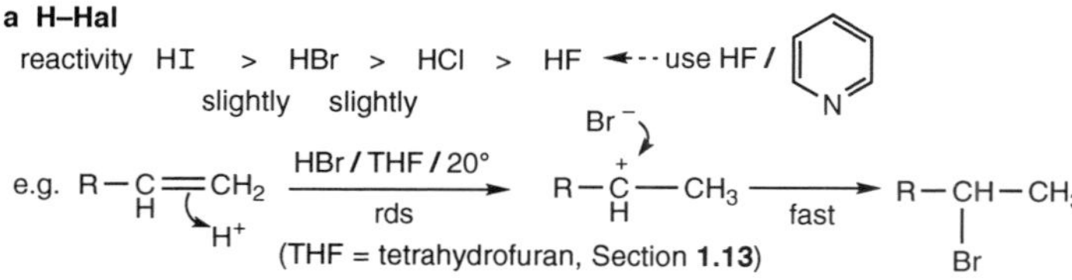

b H–OH (see also **3a,b**)

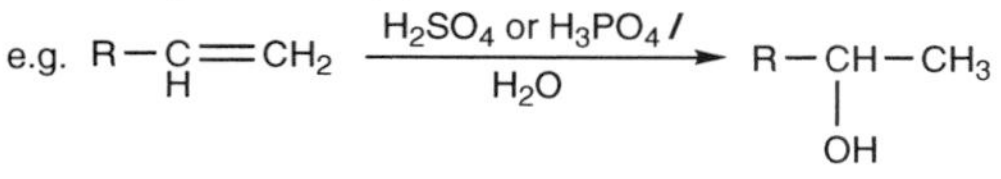

e.g. R—C=CH$_2$ $\xrightarrow[\text{H}_2\text{O}]{\text{H}_2\text{SO}_4 \text{ or H}_3\text{PO}_4 /}$ R—CH—CH$_3$
        | |
        H   OH

2 Stepwise E⁺ addition, bridged intermediate

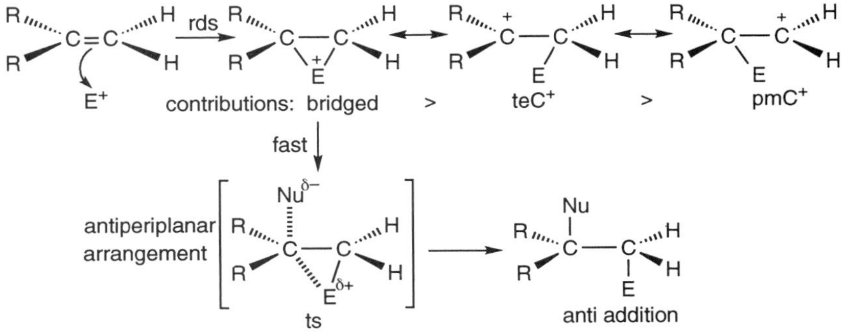

contributions: bridged > teC⁺ > pmC⁺

fast ↓

antiperiplanar
arrangement

ts

anti addition

a Hal₂

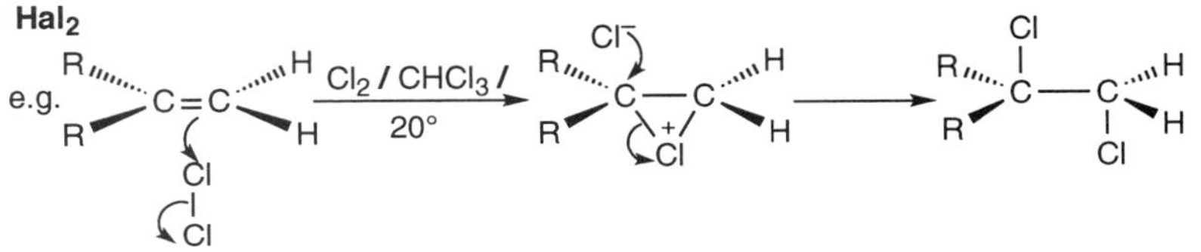

e.g.

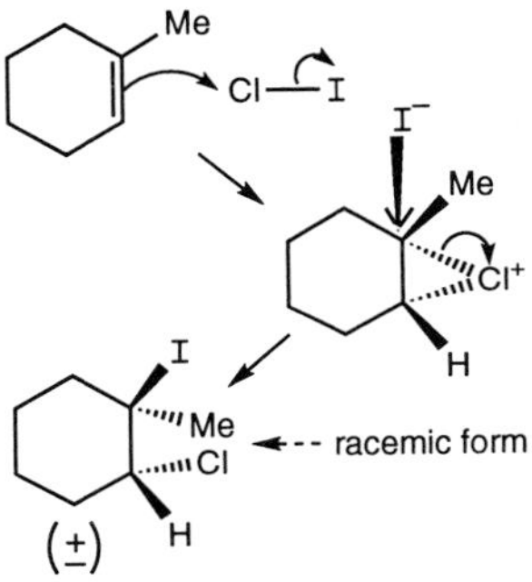

ΔH_o (kJmol⁻¹) for additions   −548   −170   −109   0

reactivity   F₂ > Cl₂ > Br₂ > I₂

too reactive, not useful   useful   does *not* add to simple alkenes

(±) racemic form

b HO–Hal

limited to Br (most useful) and Cl
solutions containing HO–Br [*hypobromous acid*, oxobromic(I) acid]
in equilibrium with its protonated form generated from Br₂ or NBS (see below)

$$Br_2 + H_2O \rightleftharpoons HBr + HO\text{–}Br \rightleftharpoons Br^- + Br\text{–}O^+H_2$$
strong acid   weak acid

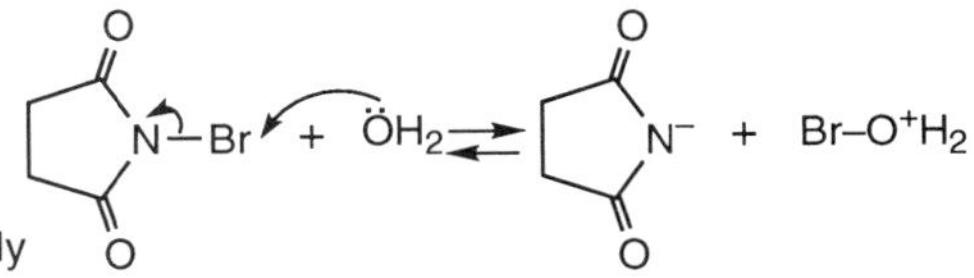

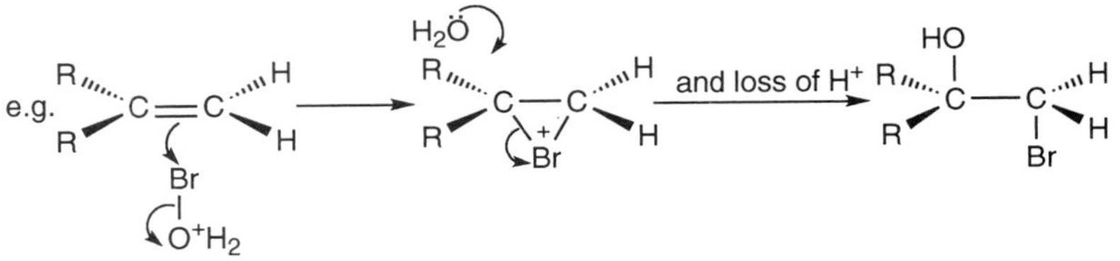

relative rates of addition of Br₂:

ethene 1   propene 61
(*E*) and (*Z*) but–2–ene ~2100
2–methylpropene 5400
2,3–dimethylbut–2–ene 1800x10³

N–bromosuccinimide,
    abbreviation NBS
N–bromobutanimide

very useful reagent, commercially
available

e.g.

and loss of H⁺

3 Concerted E^+ addition (must be syn addition)

a Hydroboration

(H C Brown, 1956 onwards)

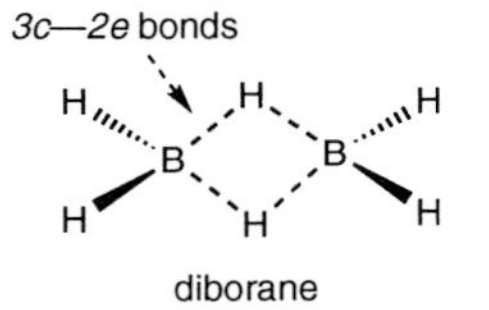
3c—2e bonds
diborane

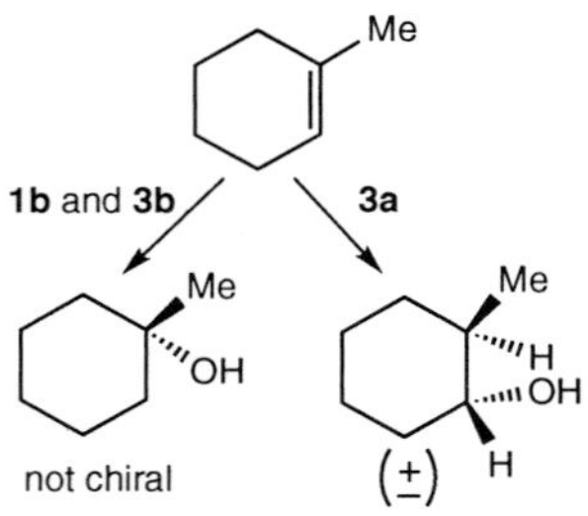
Me
1b and 3b
3a
Me
OH
Me
H
OH
H
not chiral
(±)

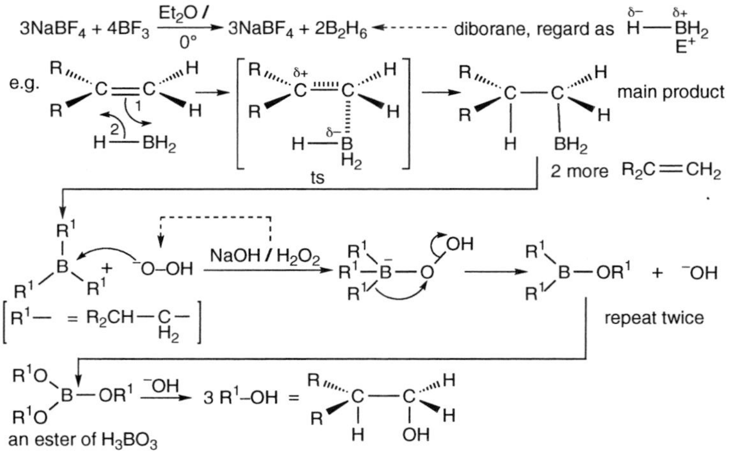
3NaBF4 + 4BF3
Et2O / 0°
3NaBF4 + 2B2H6
diborane, regard as H—BH2
δ− δ+
E+
e.g.
ts
main product
2 more R2C=CH2
NaOH / H2O2
repeat twice
[R1— = R2CH—CH2—]
an ester of H3BO3
3 R1–OH =

b $Hg(OAc)_2$ mercuric acetate [mercury(II) ethanoate]

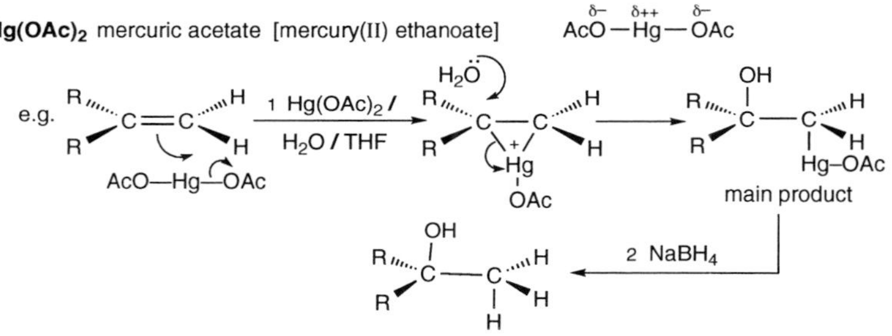
δ− δ++ δ−
AcO—Hg—OAc
e.g.
AcO—Hg—OAc
1 Hg(OAc)2 /
H2O / THF
main product
Hg–OAc
2 NaBH4

c Peroxyacids

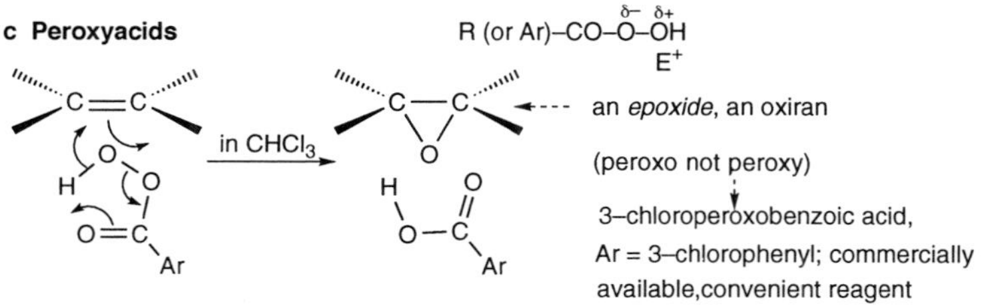
R (or Ar)–CO–O–OH
δ− δ+
E+
in CHCl3
an epoxide, an oxiran
(peroxo not peroxy)
3–chloroperoxobenzoic acid,
Ar = 3–chlorophenyl; commercially
available,convenient reagent

d Osmium tetroxide

usual conditions at 20°

very dilute solution at 0°

generally low yield

1 OsO$_4$ / Et$_2$O

2 NaHSO$_3$ / H$_2$O

HO OH

sq OsO$_4$ / R$_3$N$^+$—O$^-$ / H$_2$O / THF → HO OH + R$_3$N + OsO$_4$

sodium periodate [iodate(VII)]

sq OsO$_4$ / NaIO$_4$ / H$_2$O / THF → C=O + O=C

e Ozone

1 O$_3$ / EtOAc / −70°

no proof of existence

overall process called *ozonolysis* → C=O + O=C (+SO$_2$)

2 add Me$_2$S, allow to warm up

ozonides (explosive, don't isolate!)

4 U· reactions

a Addition of HBr

R—C=CH$_2$ $\xrightarrow{\text{HBr / sq radical initiator / hexane}}$ R—C—CH$_2$Br
   | |
   H   H$_2$

radical initiator e.g. in **2.1Re4b**; another often used, dibenzoylperoxide

Ph—C—O—O—C—Ph $\xrightarrow{\text{in solution, even at 20°}}$ 2 Ph—C—O· → Ph· + CO$_2$

init·

initiation: init· + H—Br → init–H + Br·

propagation: step 1 Br· + H$_2$C=C—R → Br—C—C—R   H$_2$C—C—R
                              | | | |
                              H   H$_2$ H   H
                                          seC· *yes*   pmC· *no* Br

step 2 Br—C—C—R + H–Br → Br—C—C—R + Br·
            | | | | chain carrier
           H$_2$ H   H$_2$ H$_2$

ΔH$_o$(kJmol^{-1})	HF	HCl	HBr	HI
steps 1 and 2 together	−49	−40	−51	−42
step 1	−220	−74	−19	56
step 2	171	34	−32	−98

b Allylic halogenation, Hal usually Cl or Br

chain carrier
Hal + Hal·

see **2.2Pr** for e.g. involving Cl$_2$ / 500°

Good general method for bromination e.g.

cyclohexene $\xrightarrow{\text{NBS / sq radical initiator / sq HBr / CCl}_4 / \Delta}$ 3–bromocyclohex–1–ene

NBS

N–Br

N–bromosuccinimide
N–bromobutanimide

c Addition of H$_2$ **4.2Pr1a** (included here although not U· in usual sense)

Most of the reactions are additions. Earlier we saw that eliminations are favoured by high temperature, because this takes advantage of the favourable ΔS_o change(**2.1Ge2d**). Additions, generally two species giving one, are entropically unfavourable; they are usually conducted at 20° to minimise this factor.

The additions to alkenes are divided between those involving electrophilic(**Re1,2** and **3**) and radical(**Re4a**) reagents. Electrophilic additions are further divided into Stepwise in which the intermediate is not bridged(**Re1**), Stepwise with a bridged intermediate(**Re2**) and Concerted(**Re3**). In all these there two important features, viz., the *orientation* and the *stereochemistry* of the reactions.

The faces of simple alkenes are equivalent (scheme). 50% of electrophiles or U· add from the top, 50% from the bottom. For simplification only one mode (from the bottom) is shown for the individual reactions of **5Re**. In many places unsymmetric aliphatic alkenes are used to signify the orientation. Unfortunately even with these the products of syn and anti addition are identical because there is free rotation about the products' CC single bonds. With cyclic systems rotation is not possible and products from syn and anti addition are different. The stereochemical features are shown clearly in the scheme, and it is good practice to work through all the reactions with, say, 1–methylcyclopentene.

In **Re1**(stepwise, intermediate not bridged) preference for a certain orientation (regiospecificity) is decided in the first step (scheme) but both syn and anti addition occur. All HHal add; with HBr it is essential to exclude strong irradiation and certain impurities which promote radical addition. H^+ is included here in the nonbridging category.($^{\#}$Later courses, the possibility of H bridges.) The results with 2–methylpropene (margin) establish kinetic control, as would be expected from the general mechanism.

With a bridged intermediate (**Re2**) the orientation is decided in the second step and anti addition occurs. The bridged canonical by itself does not allow prediction of the orientation. However, of the two C^+ canonicals that with tertiary C^+ is the more important. This is equivalent to saying that breaking the tertiary C–E bond requires less energy than breaking the primaryC–E bond. Thus the nucleophile attacks the tertiary C^+; this factor outweighs the purely steric preference of the nucleophile to attack the primary C. The favourable antiperiplanar Nu–C–C–E arrangement in the transition state leads to anti addition. In **Re2a** the lack of addition by iodine is a matter of bond energies; organic theory would not account for this. The simplest explanation of the rates of Br_2 addition (margin) is that the intermediate carbocation formed in the rate-determining step is stabilised by attachment of R groups.

The discovery of hydroboration(**Re3a**) is a milestone in the development of synthetic chemistry. Many transformations regarded previously as mere paper chemistry became going concerns. Only one application is given here; later courses will include many more. Diborane (b.p.12°) is toxic, so the solution in which it is generated is generally used directly. It is an excellent electrophile and (see later) a reducing agent. The addition is concerted but driven by the electron deficiency of boron. Thus arrow 1 (scheme) may be regarded as marginally ahead of arrow 2. The transition state involves the more stable of the two possible incipient carbocations, and boron becomes attached to the less substituted C. A steric effect, leading to the same outcome, must also be operating because bulky boranes HBR_2 show remarkably high regiospecificity. **Re3b** and **Re1b** give the same alcohol; **3b** is the preferred method in the laboratory. **3a** and **3b** give *isomeric alcohols*, the OH being added at the *less* and the *more* substituted C respectively.

Osmium tetroxide(**3d**) is an excellent reagent for hydroxylating double bonds. It is toxic (care necessary) and, for large scale work, prohibitively expensive. The amount required for 50g of but–2–ene would cost £15,000! Ingenious methods for overcoming this impediment include those (scheme) for hydroxylating and splitting double bonds. Ozone, for long regarded as a vicious indiscriminate reagent, has been developed into a good alternative method for bond fission.

Radical addition of HHal(**4a**) is restricted to HBr. Although the ΔH_o values are similar for all the HHal only HBr has reasonable values for *both* the propagation steps. The orientation is conventionally regarded as reflecting the possible C radical intermediates. [In the example shown the secondary C· is more stable than the primary C· (Section 1.9).] A steric effect, easier access to the primary C, may also be important. Radical and ionic(**1a**) addition of HBr give *isomeric bromides*, the Br being added at the *less* and the *more* substituted C respectively. At 20° radical addition occurs by both syn and anti modes. ([#]At –80° anti addition is favoured, later courses.)

Allylic chlorination (**4b**), via the stabilised allyl radical(**2.2Ge**), occurs at high temperature. Under these conditions addition would not be expected. *N*–Bromosuccinimide in dry solutions is very convenient for allylic bromination(**4b**); different conditions apply in the earlier, less useful, application(**2b**). The crucial feature of the mechanism ([#]later courses) is that Br_2 is present in *very low steady–state* (unchanging) *concentration*.

6 Alkynes

There is a close resemblance between alkynes and alkenes; the common features are not rehearsed here. Thus many reactions of alkynes, even important ones, have been omitted because their outcome is readily predicted from alkene chemistry. Attention is concentrated on the *differences* between the groups, especially those affecting synthetic work. The CC double bond is usually created from saturated intermediates by elimination. However, with the alkynes the situation is different: the simplest member, ethene, is used widely in building up the higher members and other compounds containing the triple bond. [Ethyne, *acetylene*, is produced cheaply by the industrial process of heating a mixture of methane and steam at 1400°.]

e.g. of Pr2

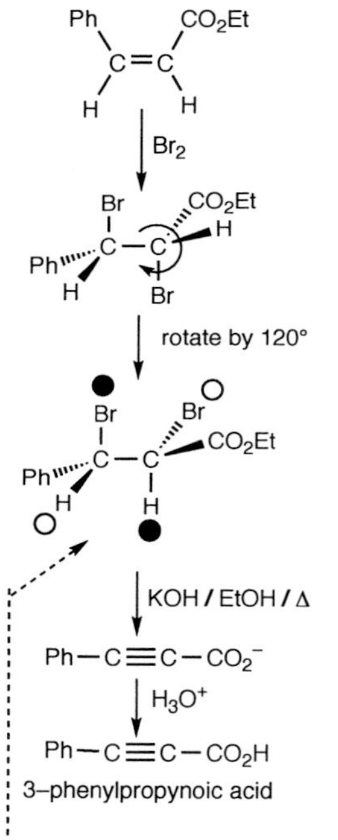

3-phenylpropynoic acid

● and ○ pairs antiperiplanar

6Ge

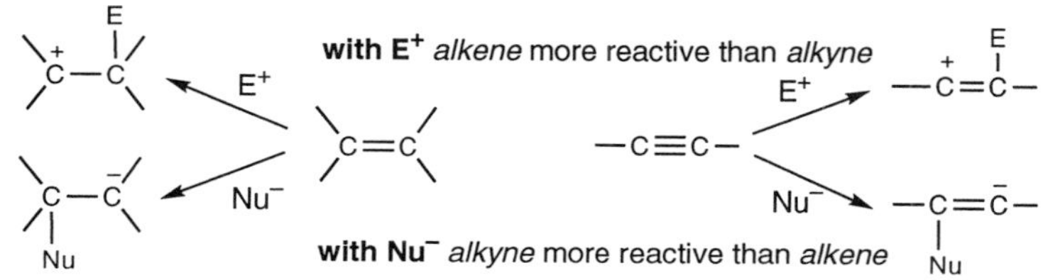

with E+ *alkene* more reactive than *alkyne*

with Nu− *alkyne* more reactive than *alkene*

At first sight it is surprising that alkynes (4 electrons in a π orbital) are less reactive than alkenes (2 such electrons) to E+ (**6Ge**). With nucleophiles the relative reactivity is reversed. Simple alkenes do not undergo nucleophilic addition; simple alkynes undergo very few such additions. These differences may be rationalised by considering the ions formed in the additions. The alkene carbanion is more stable than the alkane carbanion; the alkene carbocation is probably less stable than the alkane carbocation. (See pK_a values in **5Ge1**, the discussion in Section 1.9 and a possible qualification in **2.3Ge**.)

The main preparative routes are in **6Pr**.

6Pr

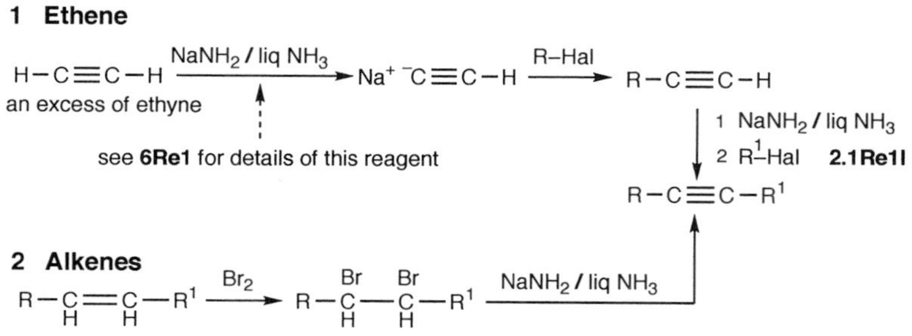

1 Ethene

see **6Re1** for details of this reagent

2 Alkenes

5Re2a

3 1,2–diketones

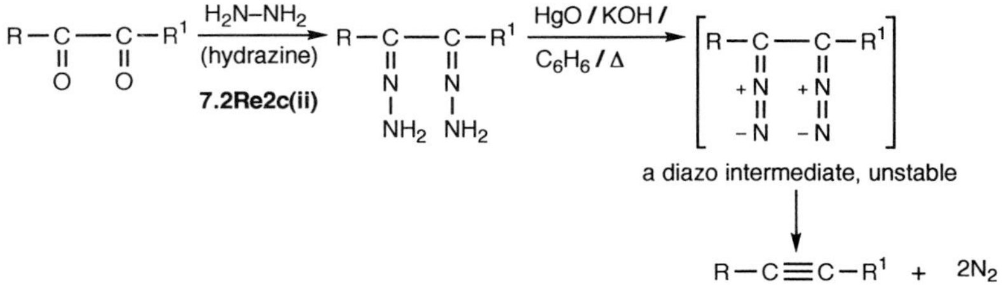

$$R-\underset{\underset{O}{\|}}{C}-\underset{\underset{O}{\|}}{C}-R^1 \xrightarrow[\substack{\text{(hydrazine)}\\ \textbf{7.2Re2c(ii)}}]{H_2N-NH_2} R-\underset{\underset{\underset{NH_2}{|}}{N}}{C}-\underset{\underset{\underset{NH_2}{|}}{N}}{C}-R^1 \xrightarrow[C_6H_6 / \Delta]{HgO / KOH /} \left[R-\underset{\substack{\|\\ +N\\ \|\\ -N}}{C}-\underset{\substack{\|\\ +N\\ \|\\ -N}}{C}-R^1 \right]$$

a diazo intermediate, unstable

$$R-C\equiv C-R^1 \;+\; 2N_2$$

As explained earlier (**2.1Re**), high yields in reactions such as **Pr1** are obtained only with primary alkyl halides. In both **1** and **2** the products must not be allowed to remain in contact with an excess of NaNH$_2$ (see **Re4**). Strained cycloalkynes were prepared by **Re3**; the drawback is that preparation of the starting materials, 1,2–diketones, involves several stages.

6Re covers reactions which have no close parallel in alkene chemistry.

6Re

1 Reduction

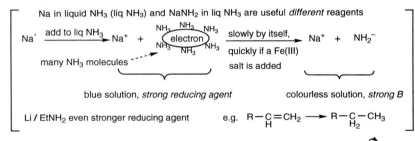

2 E⁺ addn in presence of Hg(II) compounds

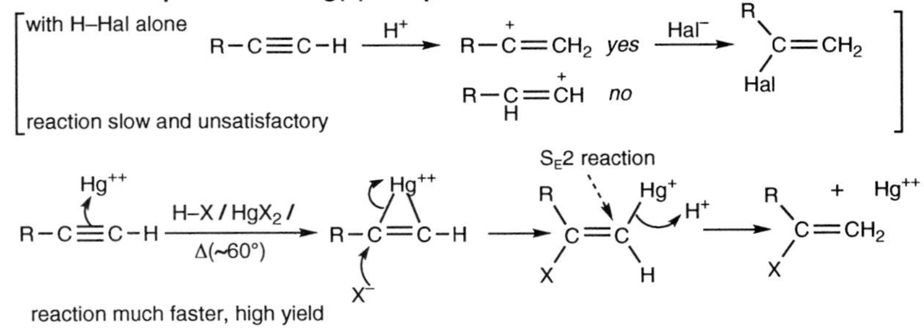

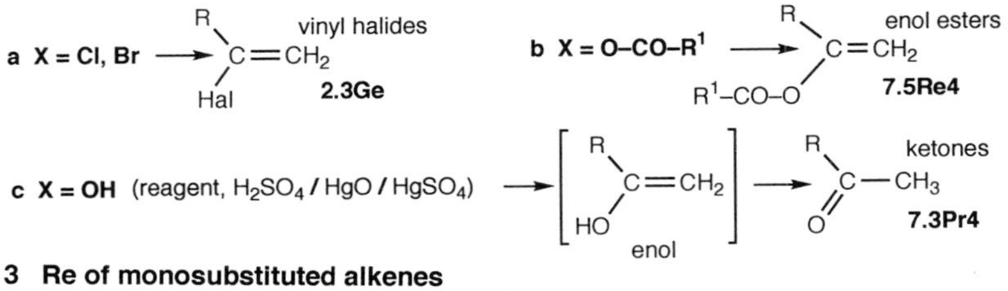

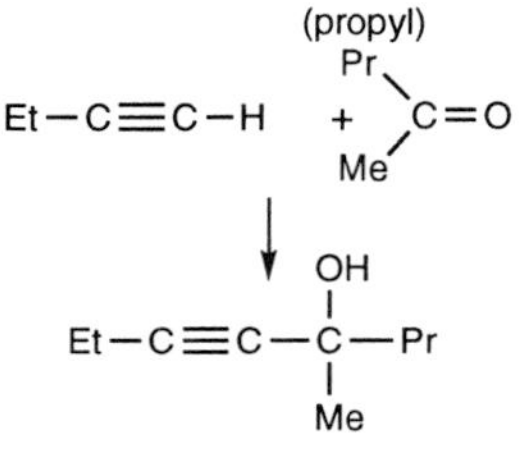

a X = Cl, Br ⟶ vinyl halides **2.3Ge**

R\C=CH₂ / Hal

b X = O-CO-R¹ ⟶ enol esters **7.5Re4**

R\C=CH₂ / R¹-CO-O

c X = OH (reagent, H₂SO₄ / HgO / HgSO₄) ⟶ [R\C=CH₂ / HO] enol ⟶ R\C-CH₃ / O ketones **7.3Pr4**

(propyl)

Et−C≡C−H + Pr\C=O / Me

⟶

Et−C≡C−C−Pr / Me , OH

4−methyloct−5−yn−4−ol

3 Re of monosubstituted alkenes

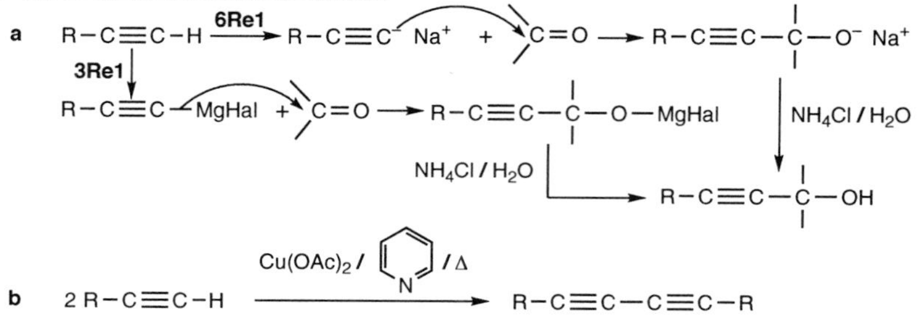

a R−C≡C−H **6Re1**⟶ R−C≡C⁻ Na⁺ + \C=O ⟶ R−C≡C−C−O⁻ Na⁺

3Re1 ↓

R−C≡C−MgHal + \C=O ⟶ R−C≡C−C−O−MgHal   NH₄Cl / H₂O

NH₄Cl / H₂O

R−C≡C−C−OH

b 2 R−C≡C−H   Cu(OAc)₂ / pyridine / Δ ⟶   R−C≡C−C≡C−R

4 Isomerization

H−C≡C−H

MeO⁻ | KOH / MeOH / Δ

H₂C=C−OMe / H

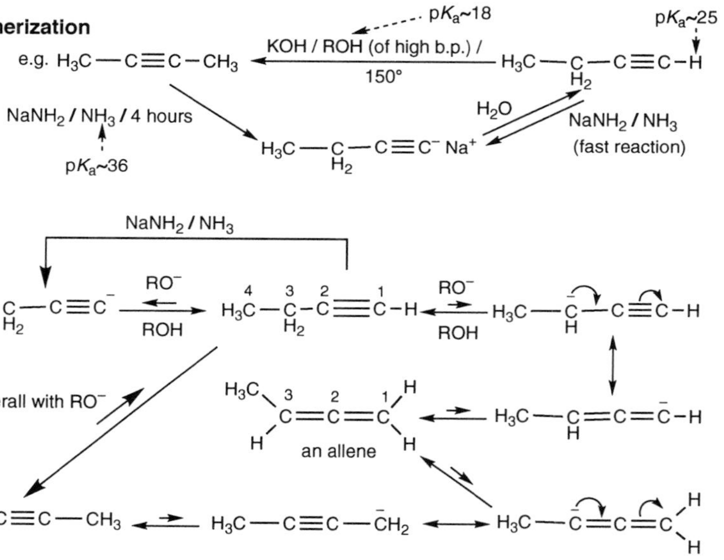

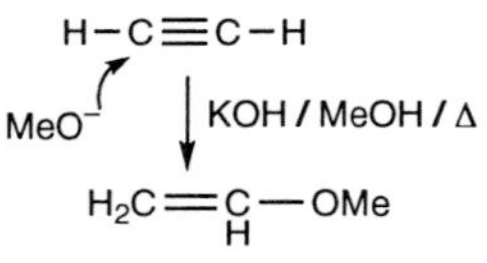

methoxyethene
methyl vinyl ether

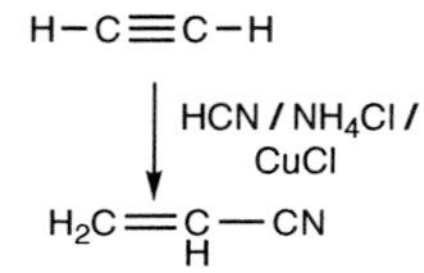

e.g. H₃C−C≡C−CH₃   ⟵ KOH / ROH (of high b.p.) / 150° ⟶ H₃C−C−C≡C−H / H₂   pKₐ~18 ... pKₐ~25

NaNH₂ / NH₃ / 4 hours ↓ pKₐ~36

H₃C−C≡C⁻ Na⁺ / H₂   H₂O NaNH₂ / NH₃ (fast reaction)

H−C≡C−H

| HCN / NH₄Cl / CuCl

H₂C=C−CN / H

propenenitrile
acrylonitrile

NaNH₂ / NH₃

H₃C−C−C≡C⁻ / H₂   ⟵ RO⁻ / ROH ⟶   H₃C−C−C≡C−H / H₂ (4 3 2 1)   ⟵ RO⁻ / ROH ⟶   H₃C−C−C≡C−H / H

overall with RO⁻

H₃C\C=C=C / H , H (3 2 1) an allene   ⟶   H₃C−C=C=C−H / H

H₃C−C≡C−CH₃   ⟵   H₃C−C≡C−CH₂   ⟶   H₃C−C=C=C / H , H

Hydrogenation of an alkyne with a poisoned catalyst(**6Pr1**) stops at the alkene stage. The product is, as expected from the syn addition shown in **4.1Pr1a**, the cis (*Z*) isomer. The alternative reduction involves sodium in liquid ammonia (b.p.−33°). [Liquid NH₃ is used in a Dewar flask, the laboratory equivalent of a thermos flask, which is confined to the fume cupboard!] The virtue of liquid NH₃ is its extraordinary ability to solvate 'free' electrons. Thus addition of Na gives Na⁺ and what may be regarded as a solvated electron. In the radical–ion formed by addition of the electron to the triple bond the alkyl groups adopt the trans arrangement (scheme) and the product is the trans (*E*) isomer. Reductions of triple bonds are important tools in syntheses: *an alkyne gives a cis or a trans alkene in high yield*. It is important to distinguish clearly between Na and NaNH₂ in liquid NH₃ (scheme).

The electrophilic additions (**Re2**) are general for CC triple bonds. However they are illustrated with monosubstituted alkynes to show the orientation. Addition, slow with electrophilic reagents, is greatly facilitated by Hg(II) salts. The mechanisms suggested, interpreted along the lines discussed under **5Re1** and **2**, account for the observed orientations ($^{\#}S_E2$ reaction, later courses). **Re2c** forms an enol, the unstable isomer of a ketone.

Re3 involves the terminal alkyne H ($pK_a \sim 25$). The Na and Mg derivatives are useful nucleophilic reagents. Addition to ketones as illustrated, to build up more complex molecules, is a frequent ploy in synthesis. The oxidation **Re3b** proceeds in high yield; its mechanism remains obscure.

Isomerisations(**Re4**) are based on 2 features: a disubstituted alkyne is more stable than a monosubstituted alkyne, and an alkyne H is much more acidic than an alkane H(**5Ge1**). Relevant pK_a values are shown in the scheme. RO^- is a weak base. With the monosubstituted alkyne it forms only *very small amounts* of negative ions by removing the terminal alkyne H or an H of the CH_2 attached to the triple bond. Equilibria are set up with an allene intermediate (scheme) which may lose H from C_1 or C_3. Loss from C_1 results in reversion to the original alkyne; loss from C_3 leads on to the more stable isomer. NH_2^- is a very strong base. Equilibria are again set up. However an alkyne H is completely removed by NH_2^-. Thus, as the terminal alkyne is formed it is trapped as its Na salt. Both reactions are under thermodynamic control.

Two reactions of industrial importance (margin) appear to be restricted to ethyne. They do not occur with other simple alkynes, not even with propyne. That giving methoxyethene is remarkable in exemplifying *nucleophilic addition to an alkyne*.

7   Aldehydes and ketones

The CO double bond (carbonyl group) is the commonest group in organic chemistry. Many tomes have been devoted entirely to this functional group. Our task is to drive a straight course through carbonyl chemistry, glancing to left and right but not digressing along any of the innumerable side roads. For this purpose it is helpful to divide the subject into manageable sections, viz.,

Section 7.1 The carbonyl group, keto and enol forms
Section 7.2 Oxidation and reduction of carbonyl compounds
Section 7.3 Preparations
Section 7.4 Reactions of keto forms
Section 7.5 Reactions of enol forms and enolate anions
Section 7.6 Aldol and related condensations

7.1   The carbonyl group, keto and enol forms

K_1 of equilibria, pK_a of **H**

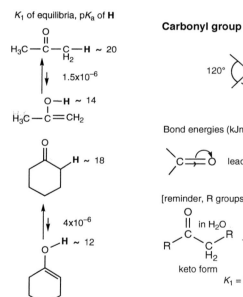

7.1Ge

Carbonyl group

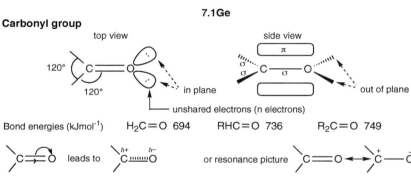

Bond energies (kJmol^{-1}) $H_2C=O$ 694 $RHC=O$ 736 $R_2C=O$ 749

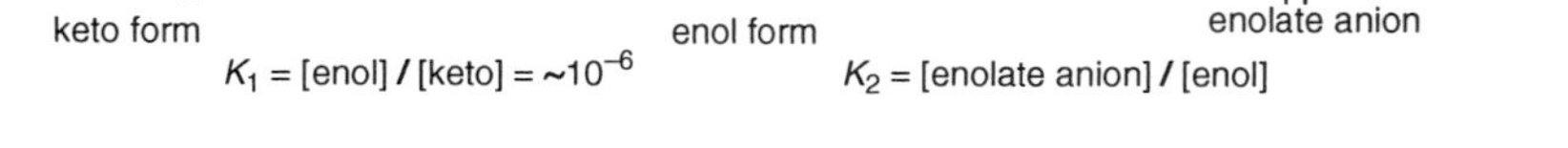

$$\begin{array}{ccc} \diagdown \\ \diagup \end{array}C \!\!\rightarrow\!\! O \quad \text{leads to} \quad \overset{\delta+}{C}\!\cdots\!\overset{\delta-}{O} \quad \text{or resonance picture} \quad C\!=\!O \longleftrightarrow \overset{+}{C}\!-\!\overset{-}{O}$$

[reminder, R groups may be identical or different]

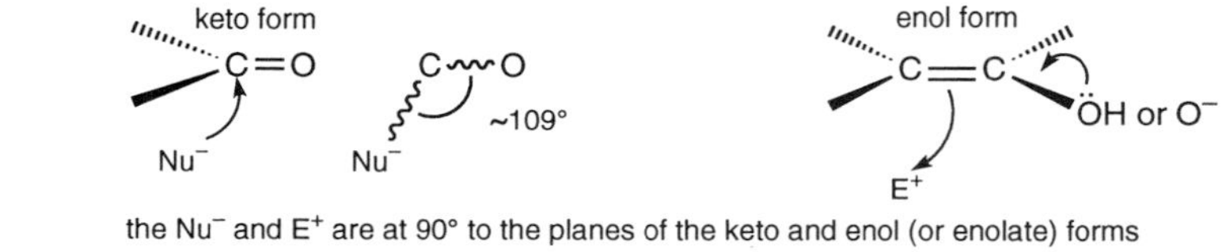

keto form $\xrightarrow[\text{in } H_2O]{H_3O^+ \text{ or } OH^-}$ enol form $\xrightleftharpoons{OH^-}$ enolate anion

$K_1 = \text{[enol] / [keto]} = \sim10^{-6}$ $K_2 = \text{[enolate anion] / [enol]}$

Fundamental tendencies

keto form   enol form

$\sim109°$

Nu^- Nu^- E^+

the Nu$^-$ and E$^+$ are at 90° to the planes of the keto and enol (or enolate) forms

Scheme **7.1Ge** shows the main features. The polarisation can be expressed in terms of electronic effects or of resonance: the outcome is the same. Electrophiles (notably H⁺) add to O$^{\delta-}$, but the addition is reversible. The C$^{\delta+}$ centre is more important, as discussed later. Carbonyl compounds exhibit *tautomerism*. (A tautomeric compound, under suitable conditions, consists of an equilibrium mixture of two or more structural isomers. Tautomers are real species; canonicals are not.) Simple carbonyl compounds contain only very small amounts of the enolic forms (margin). In a solution free from impurities *interconversion between keto and enol forms is slow*. Adding acids or bases *increases the rate of interconversion but the position of the equilibrium (K_1) is not changed*. With a base there is a second equilibrium, between enol and enolate–anion; *the position of this equilibrium (K_2) does depend on [base]*. Mechanisms for these interconversions are in **7.5Re1a**. The complexity of carbonyl chemistry arises from a duality: *the keto form reacts with nucleophiles, the enol form with electrophiles*. In both types of reaction the outcome may depend on rate factors (kinetic control) or on the position of equilibria (thermodynamic control) (Section 1.5).

7.2 Oxidation and reduction of carbonyl compounds

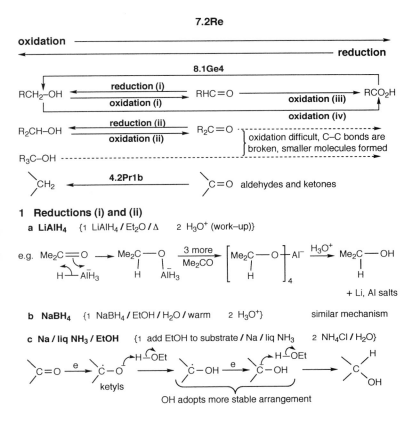

1 Reductions (i) and (ii)

a LiAlH₄ {1 LiAlH₄ / Et₂O / Δ   2 H₃O⁺ (work–up)}

b NaBH₄ {1 NaBH₄ / EtOH / H₂O / warm   2 H₃O⁺}   similar mechanism

c Na / liq NH₃ / EtOH {1 add EtOH to substrate / Na / liq NH₃   2 NH₄Cl / H₂O}

2 Oxidations (ii), (iii) and (iv)

a chromic acid [Cr(vi)]

$$R_2CH\text{–}OH \xrightarrow{\quad CrO_3 / H_2SO_4 / H_2O / Me_2CO \text{ (solvent)} \quad} R_2C{=}O$$

$$R_2CH\text{–}OH \underset{\text{fast}}{\overset{H_2CrO_4}{\rightleftharpoons}} \quad \xrightarrow{\text{rls}} \quad R_2C{=}O \; + \; \underset{O}{Cr}\text{–OH}$$

esters of chromic acid

$$RCH_2\text{–}OH \xrightarrow{\quad H_2CrO_4 \quad} RHC{=}O \xrightarrow{\qquad} RCO_2H$$

b AgNO$_3$ / NH$_3$ / H$_2$O

$$RHC{=}O \; + \; Ag^+(NH_3)_2 \longrightarrow RCO_2H \; + \; Ag$$

(as silver mirror under controlled conditions)

3 Oxidation (i)

CrO$_3$ / pyridine / CH$_2$Cl$_2$ reagents must be *dry*

$$RCH_2\text{–}OH \longrightarrow RHC{=}O$$

if *H$_2$O* present

$$RHC{=}O \rightleftharpoons RC\underset{OH}{\overset{H}{\text{–}OH}} \longrightarrow RCO_2H$$

7.4Re2b(i)

4 Reduction (ii), Oxidation (ii)

$$R_2CH\text{–}OH \underset{\big[\,Al{+}O\text{–}CHMe_2\,\big]_3 / Me_2CHOH / \Delta}{\overset{\big[\,Al{+}O\text{–}CMe_3\,\big]_3 / Me_2CO / \Delta}{\rightleftharpoons}} R_2C{=}O$$

‹----- Oppenauer oxidation

‹----- Meerwein–Ponndorf reduction

overall $R_2CH\text{–}OH \; + \; Me_2CO \rightleftharpoons R_2C{=}O \; + \; Me_2CHOH$

excess drives reaction to right   excess drives reaction to left

#5 Oxidations (i) and (ii)

$$\underset{H}{\overset{OH}{C}} + Me_2S{=}O \xrightarrow[\text{DMSO}]{\text{suitable } E^+} C{=}O \; + \; Me_2S$$

Pfitzner–Moffatt oxidation

mechanism starts here

e.g. $Me_2\overset{+}{S}\text{–}O\text{–}Ac \; + \; Ac\text{–}O^- \longleftarrow Me_2S{=}O \; + \; Ac\text{–}O\text{–}Ac$

ethanoic anhydride

$$\underset{H}{\overset{\ddot{O}H}{C}} \longrightarrow \underset{H}{\overset{O\text{–}\overset{+}{S}\text{–}CH_2}{C}}_{Me} \longrightarrow \overset{O}{\underset{H}{C}}\overset{+}{\underset{CH_2}{S}}\text{–}Me \longrightarrow C{=}O \; + \; Me_2S$$

6 Oxidation of allyl and benzyl alcohols

$$\underset{\underset{OH}{\overset{|}{C}}}{C{=}C}\text{–H} \xrightarrow[\sim 12 \text{ hours at } 20°]{MnO_2 / \text{hexane} / \text{stir}} \underset{\underset{C{=}O}{\overset{|}{C}}}{C{=}C}$$

$$\underset{OH}{\overset{Ar \quad H}{C}} \longrightarrow \overset{Ar}{C{=}O}$$

hydride transfer to
carbonyl group
(recalls **3Re3** product **V**)

B(O–CO–Et)$_3$

B$_2$H$_6$ | Et–CO$_2$H /
Δ

NaBH$_4$

The transformations are labelled so that the scope of the methods can be defined. Aldehydes prefer to go to acids rather than to alcohols: aldehydes are reducing agents, ketones are not. Lithium aluminium hydride (**Re1a**) is an excellent versatile reducing agent. Even the simplified mechanism shown brings out a crucial feature, that the process involves coordination to the O and is not merely addition of an H^- to the C. Sodium hydride, which does not coordinate, is a strong base but not a reducing agent. Sodium borohydride (**1b**) is a milder more selective agent. There is by now a whole family of hydride reagents, many of which are selective in reducing only certain groups. $LiAlH_4$ and $NaBH_4$ add irreversibly, and when the two faces of the CO group are not equivalent the product formed depends on the rates of the different approaches (kinetic control). Reduction of aldehydes and ketones by diborane (**5Re3a;8.1Ge4**) is fairly slow. The example in the margin shows how this reagent can be used to reduce a CC double bond without affecting a CO group. In **1c** EtOH provides a relatively acidic H which protonates a ketyl quickly. If the ketyl lingers it tends to dimerise (**7.4Re1d**). Dissolving metal reduction usually gives the more stable alcohol (thermodynamic control).

Chromic acid(**Re2a**) is excellent for preparing ketones, but aldehydes are oxidised further to carboxylic acids. Modification **3** does stop at the aldehyde. Bonding with pyridine protects the aldehyde. Dry reagents are essential (scheme). At first sight the material in **4** appears discouragingly complicated, but the chemistry is in fact readily understandable. Aluminium alkoxides can transfer a hydride to the carbonyl group. The process is reversible (margin). Thus by adroit choice of alkoxides and solvents the reaction can be driven in both directions, as an oxidation or a reduction. [#]The chemistry in **5** *is* complicated. For simplicity ethanoic anhydride *(acetic anhydride)* is shown as the electrophile but others, for example carbodiimides, are more effective. Method **6** is selective for allyl and benzyl alcohols; simple alcohols are not oxidised by this mild treatment.

7.3 Preparations

The main routes to carbonyl compounds are in scheme **7.3Pr**.

7.3Pr

1 **R–OH** 7.2Ge   2 **R–M** 3Re2 d–h

3 **Alkenes** 5Re3d,e   4 **Alkynes** 6Re2c

5 *Acetoacetic ester,* **Ethyl 3-oxobutanoate** ┈┈► Pr in **7.6Re2a; 8.3Re** (margin)

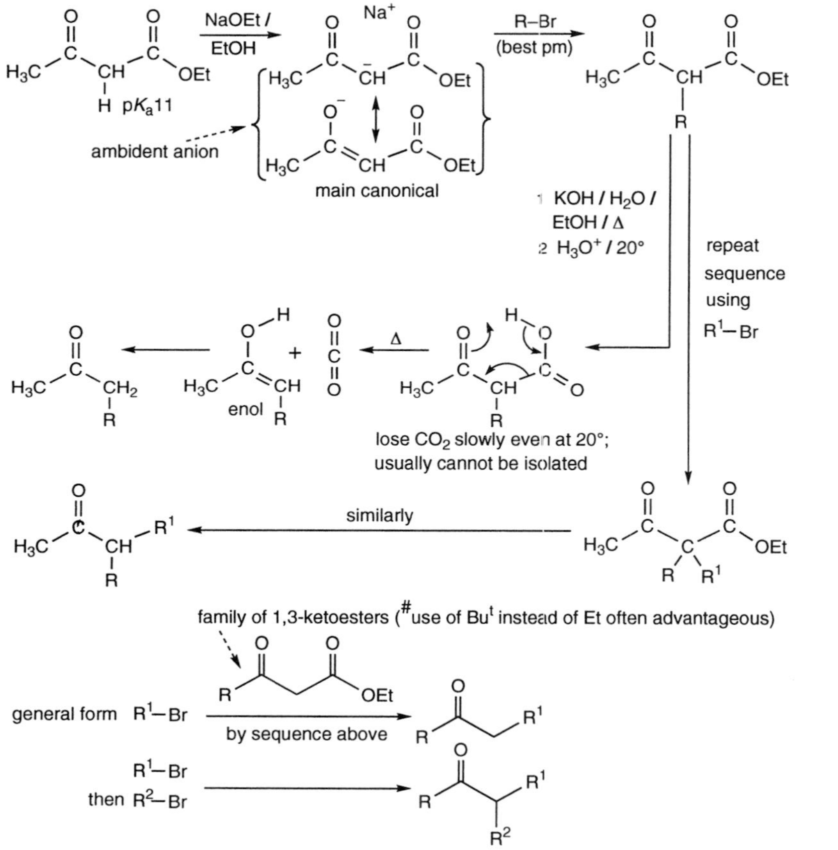

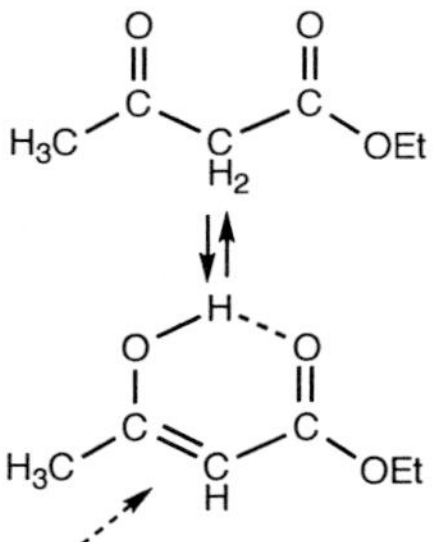

% enol

in hexane	49
in EtOH	11
neat liquid	8
in H$_2$O	0.5

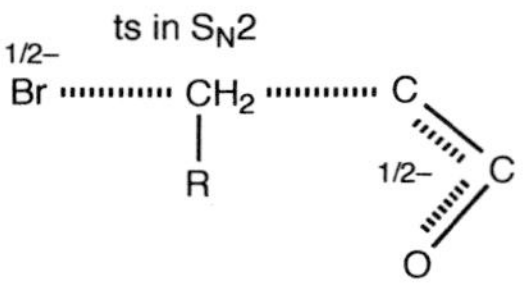

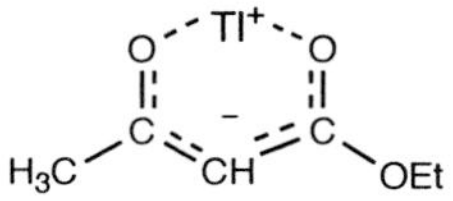

ts in S$_N$2

good with primary
and secondary R–Br

NaOEt /
EtOH

H pK$_a$11

ambident anion

main canonical

1 KOH / H$_2$O /
   EtOH / Δ
2 H$_3$O$^+$ / 20°

repeat
sequence
using
R^1–Br

R–Br
(best pm)

enol

lose CO$_2$ slowly even at 20°;
usually cannot be isolated

similarly

family of 1,3-ketoesters ($^\#$use of Bu^t instead of Et often advantageous)

general form R^1–Br

by sequence above

R^1–Br
then R^2–Br

Four of the methods have already been covered. The fifth(**Pr5**) involves
acetoacetic ester. Its properties and reactions are general for 1,3–ketoesters,
important compounds in synthetic work. The enol (margin) contains a
conjugated system; this and, to some extent, the intramolecular bonding
stabilise the enol. Thus compared with simple ketones the tautomeric
equilibrium is far more towards the enol side. With sodium ethoxide (1 mol)
acetoacetic ester (1 mol) is converted, virtually completely, into the ambident
ion shown. (Ambident ions, **2.1Rej,k.**) The question arises why does
alkylation, best with primary R–Hal, occur almost exclusively at C rather

than at O? Theories abound; none explains satisfactorily the mass of information in the literature. A useful generalisation, not without exceptions, is that S_N2 occurs at the less electronegative centre (margin). The electrophilic C of the attacking halide is almost neutral in the transition state of the S_N2 process (**2.1Ge1a**). This generalisation is a corollary of the earlier one about ambident ions (bottom of **2.1Re2**). The key reaction of 1,3–ketoacids is the facile decarboxylation (scheme) which generates ketones. Alkylations with secondary R–Hal are improved by using the thallium rather than the sodium salt (margin).

7.4   Reactions of keto forms

The reactions of keto forms with nucleophiles are in scheme **7.4Re**

7.4Re

aldehyde has more δ+ C and approach less hindered

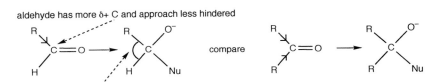

less 'back strain' created

1   Irreversible

a R–M   **3Re2a**   **b** Ylids   **5Pr1a**   **c** R—C≡C⁻ Na⁺   **6Re3a**
                                                                     R—C≡C—MgHal

d   M such as Mg, Al   (included here by considering an electron as a nucleophile)

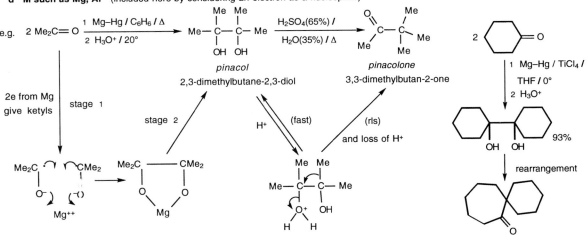

yield improved by adding TiCl₄ in stage1 (TiCl₂, a good reducing agent, is formed)

with CF$_3$CO$_3$H / Na$_2$HPO$_4$

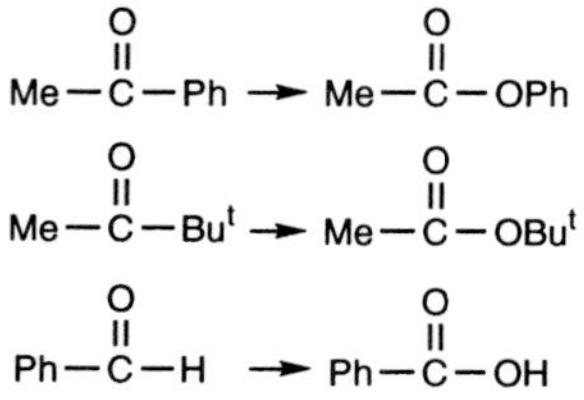

$$Me-\overset{O}{\overset{\|}{C}}-Ph \rightarrow Me-\overset{O}{\overset{\|}{C}}-OPh$$

$$Me-\overset{O}{\overset{\|}{C}}-Bu^t \rightarrow Me-\overset{O}{\overset{\|}{C}}-OBu^t$$

$$Ph-\overset{O}{\overset{\|}{C}}-H \rightarrow Ph-\overset{O}{\overset{\|}{C}}-OH$$

e R^1–CO–O–OH (peroxyacids **5Re3c**)

Baeyer–Villiger oxidation (1899)

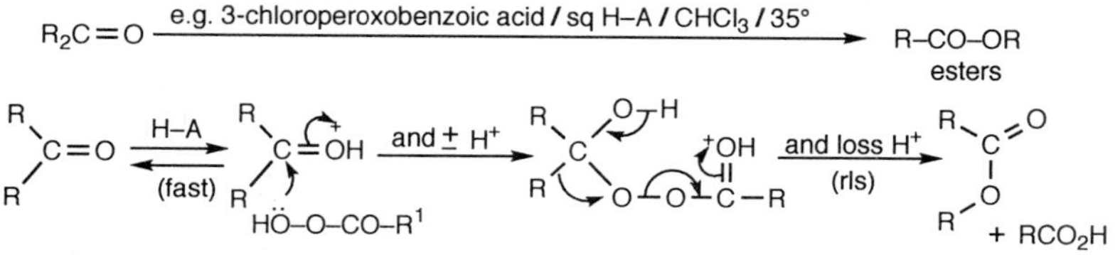

$$R_2C=O \xrightarrow{\text{e.g. 3-chloroperoxobenzoic acid / sq H–A / CHCl}_3 / 35°} \text{R–CO–OR}$$

esters

± H$^+$ indicates fast loss of H$^+$ from one site and fast addition at another site

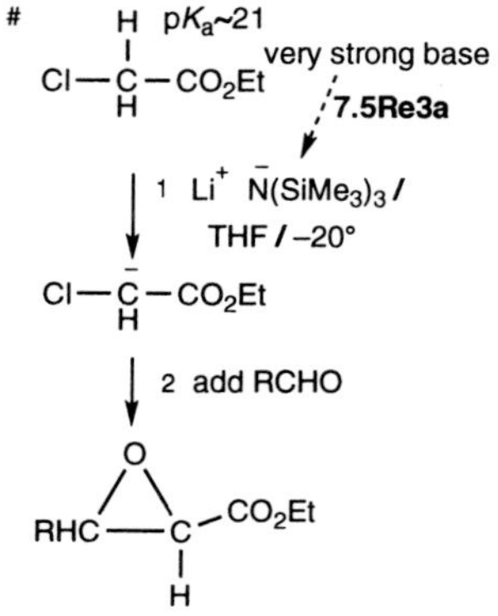

f Esters of 2-chloroacids ···· glycidic ester Darzens reaction

$\underset{\overset{|}{H}}{\overset{H \ pK_a \sim 21}{Cl-\overset{|}{C}-CO_2Et}}$ very strong base **7.5Re3a**

↓ 1 Li$^+$ $\overset{-}{N}$(SiMe$_3$)$_3$ / THF / –20°

$Cl-\overset{|}{\underset{H}{\overset{-}{C}}}-CO_2Et$

↓ 2 add RCHO

$\underset{\overset{|}{H}}{RHC-\overset{O}{\triangle}-\overset{|}{C}-CO_2Et}$

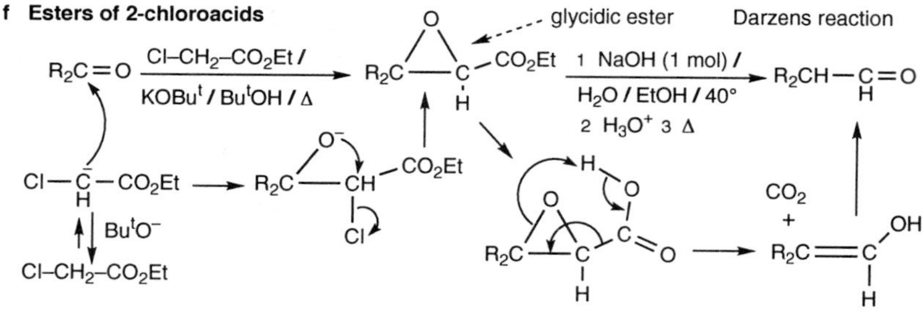

2 Reversible

nitrogen

O, N, C

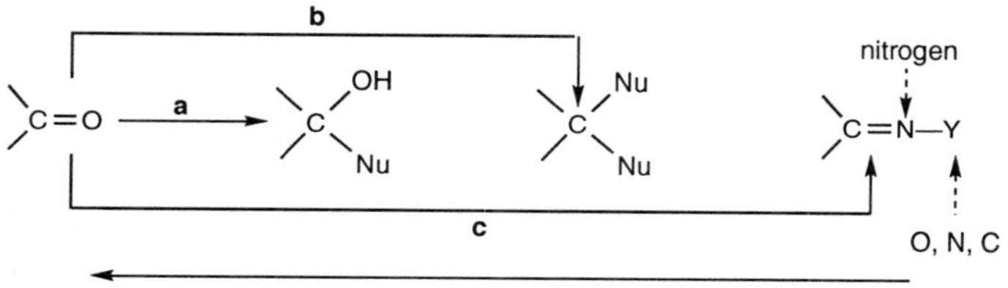

reversed with H$_3$O$^+$ / Δ if not specified otherwise

a(i) NaHSO$_3$ sodium bisulphite [trioxosulphate(v)]

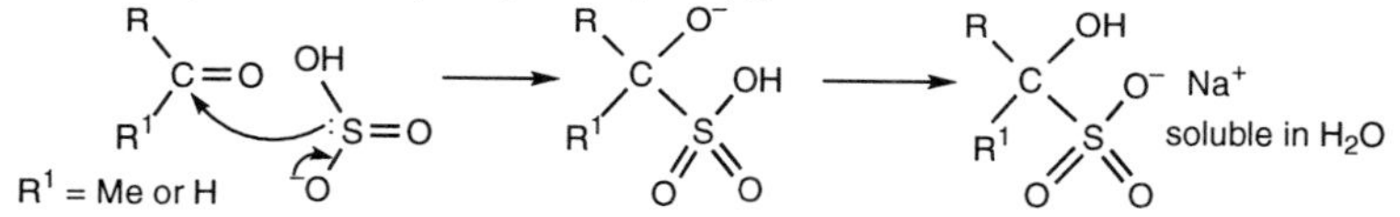

R^1 = Me or H soluble in H$_2$O

(ii) HCN ···· covalent molecule not a Nu$^-$

$$\underset{}{\overset{}{C}}=O \xrightarrow[\text{1 KCN 2 add H}_3\text{O}^+ \text{ slowly}]{\text{sq KCN / HCN or}} \underset{CN}{\overset{OH}{C}}$$

cyanohydrins
1,1-hydroxynitriles

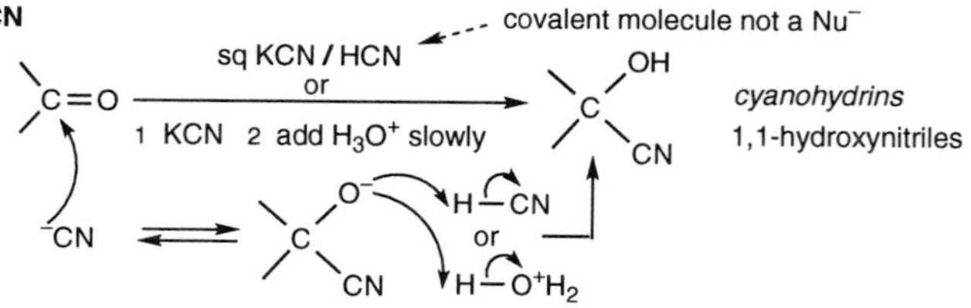

% hydrate in H$_2$O

CH$_2$O	~100
MeCHO	60
Me$_2$CO	~0
CCl$_3$CHO	100

CCl$_3$CH(OH)$_2$ is a crystalline solid

b(i) H$_2$O

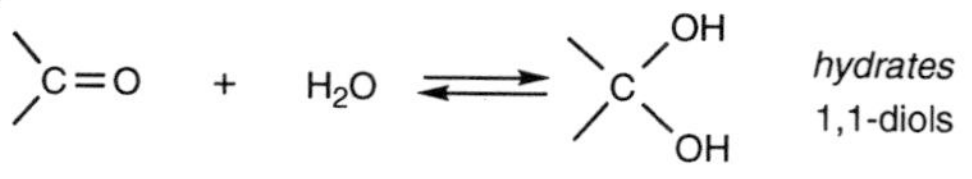

$$\underset{}{\overset{}{C}}=O \ + \ H_2O \rightleftharpoons \underset{OH}{\overset{OH}{C}}$$

hydrates
1,1-diols

(ii) ROH

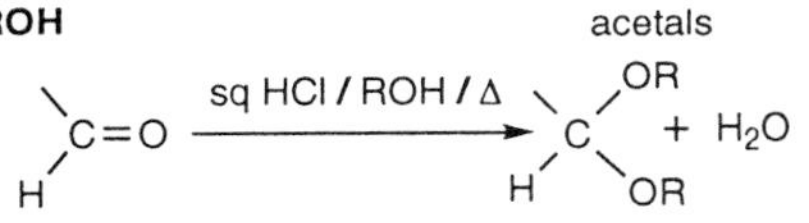

acetals

$$\ce{\underset{H}{>}C=O} \xrightarrow{\text{sq HCl / ROH / }\Delta} \ce{\underset{H}{>}C\overset{OR}{_{OR}}} + H_2O$$

all reactants *dry*, general for aldehydes with all primary ROH; ketones do not react under these conditions

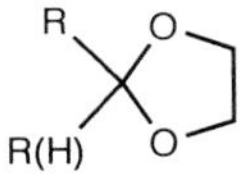

R(H) cyclic acetals in carbonyl protection

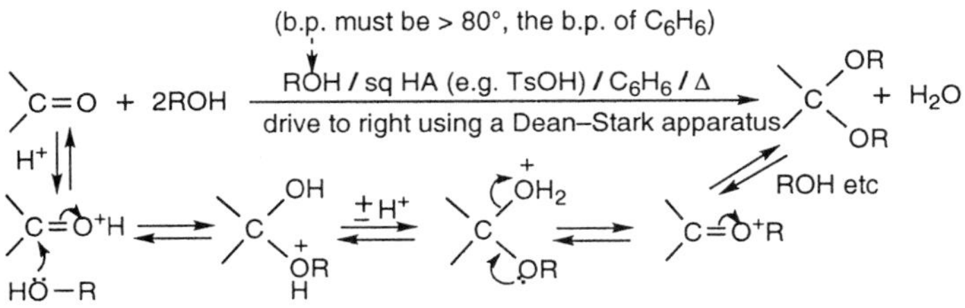

(b.p. must be > 80°, the b.p. of C_6H_6)

$$\ce{>C=O} + 2ROH \xrightarrow[\text{drive to right using a Dean–Stark apparatus}]{\text{ROH / sq HA (e.g. TsOH) / }C_6H_6\text{ / }\Delta} \ce{>C\overset{OR}{_{OR}}} + H_2O$$

ROH etc

(iii) RSH

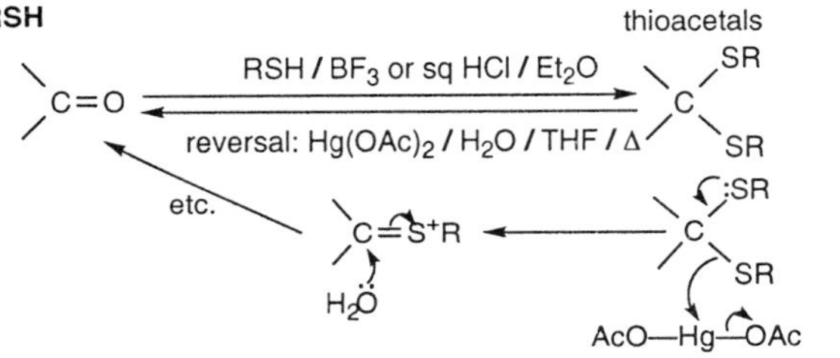

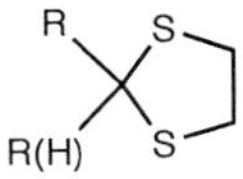

thioacetals

$$\ce{>C=O} \underset{\text{reversal: Hg(OAc)}_2\text{ / }H_2O\text{ / THF / }\Delta}{\overset{\text{RSH / BF}_3\text{ or sq HCl / Et}_2O}{\rightleftharpoons}} \ce{>C\overset{SR}{_{SR}}}$$

etc.

AcO—Hg—OAc

R(H) cyclic thioacetals in carbonyl reduction **4.2Pr1b(ii)**

c(i) H_2N–OH (hydroxylamine)

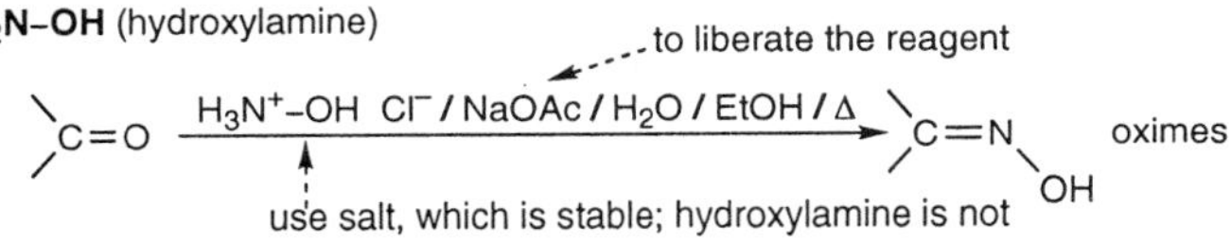

to liberate the reagent

$$\ce{>C=O} \xrightarrow{H_3N^+\text{–OH Cl}^-\text{ / NaOAc / }H_2O\text{ / EtOH / }\Delta} \ce{>C=N–OH}$$ oximes

use salt, which is stable; hydroxylamine is not

(ii) H_2N–NH_2 (hydrazine) and derivatives

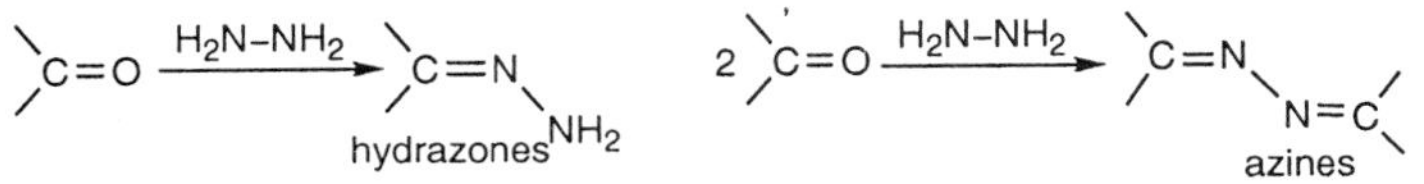

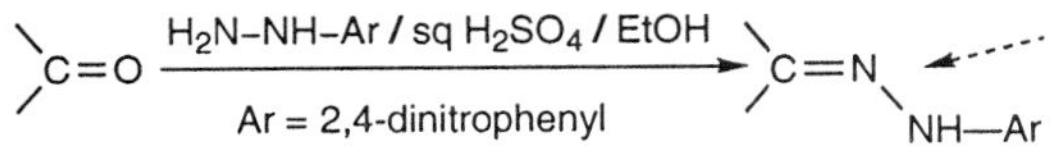

$$\ce{>C=O} \xrightarrow{H_2N-NH_2} \ce{>C=N-NH_2}$$ hydrazones

$$2\ce{>C=O} \xrightarrow{H_2N-NH_2} \ce{>C=N-N=C<}$$ azines

$$\ce{>C=O} \xrightarrow[Ar = \text{2,4-dinitrophenyl}]{H_2N-NH-Ar\text{ / sq }H_2SO_4\text{ / EtOH}} \ce{>C=N-NH-Ar}$$

yellow or red precipitate, classical test for carbonyl compounds

many others e.g.
H_2N–NH–CO–NH_2 (semicarbazide) ⟶ $\ce{>C=N-NH-CO-NH_2}$ semicarbazones
see also **4.2Pr1b(i)**

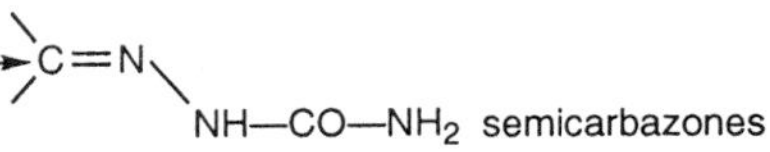

(iii) R–NH_2 and R_2NH (amines)

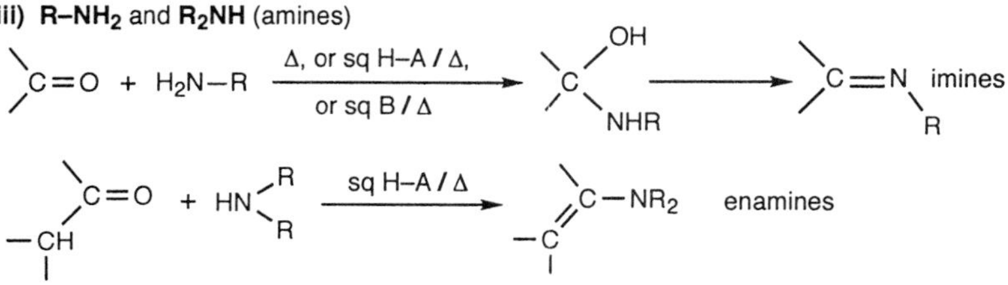

$$\ce{>C=O} + H_2N-R \xrightarrow[\text{or sq B / }\Delta]{\Delta\text{, or sq H–A / }\Delta,} \ce{>C\overset{OH}{_{NHR}}} \longrightarrow \ce{>C=N-R}$$ imines

$$\ce{>C=O} + HN\overset{R}{_{R}} \xrightarrow{\text{sq H–A / }\Delta} \ce{-C-NR_2}$$ enamines

e.g.s of stable imines

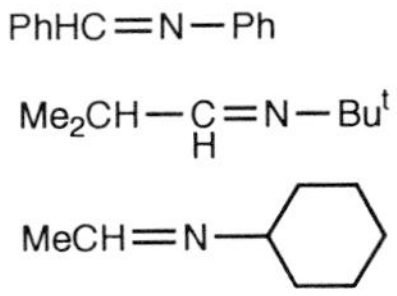

$PhHC=N-Ph$

$Me_2CH-\underset{H}{C}=N-Bu^t$

$MeCH=N-$⬡

Re2c(iii)
e.g. of enamine [#]and its use

a 1,3-diketone

Aldehydes are more reactive than ketones. Three relevant factors are shown at the top of scheme **7.4Re**. The contribution of 'back strain' (discussion in **2.1Ge2**) should not be overlooked. The higher reactivity of aldehydes is an advantage in 'unreactive situations' but there is a snag. With acids and bases aldehydes are prone to self-condensations (Section 7.6) which may lead to polymers; *thus many reactions of ketones are not applicable to aldehydes* unless the conditions are modified (see example later, **Re1f**). Reactions represented by the formulae with free bonds are general for aldehydes and ketones. Elsewhere the main restrictions are noted.

Dimerisation of ketyls(**Re1d**) was mentioned earlier(**7.2Re1c**). The pinacol–pinacolone rearrangement shown in **1d** is the archetype of countless reactions in which groups migrate to a neighbouring centre. With ketones having different R groups oxidation **Re1e** may give two products: the migratory aptitudes (margin) accord with the simple view that a group migrates as an incipient anion (i.e. with its bonding electrons). Peroxyacids oxidise aldehydes to carboxylic acids by migration of a hydride ion. Under the standard conditions of the Darzens reaction(**1f**) aldehydes undergo base induced condensations (Section 7.6). This difficulty is overcome by using all the B in a complete conversion of the 2–chloroester into its anion before the aldehyde is added (margin).

Of the **Re2** reactions many are accelerated by acids, some by acids or bases, some by bases, and a few do not require assistance. ([#]Later courses, distinction between specific and general acid and base catalysis; no distinction is attempted here.) The notion of establishing detailed mechanisms for organic reactions stems from the study of **Re2a(ii)** by Lapworth (work in 1903 at the Goldsmiths Institute, London). In the procedure involving hydrogen cyanide, cyanide ion (the nucleophile) is regenerated in the second step. The use of HCN (Prussic acid, a volatile liquid, b.p. 23°) is not for the faint hearted! Even the second procedure, with KCN, requires careful manipulation. Stable hydrates[**2b(i)**] are formed by aldehydes having strongly electron withdrawing groups (margin).

Acids promote both formation and hydrolysis of acetals[**2b(ii)**] by the common mechanism shown. The reaction is directed by removing H_2O or using an excess of H_2O. [Details of apparatus, such as Dean–Stark and Soxhlet (needed later in **7.6Re1a**), will be found in the modern practical book by L M Harwood and C J Moody, 'Experimental Organic Chemistry'.] Acetals are stable to alkali. Thioacetals[**2b(iii)**] are stable to alkali and acid; hydrolysis requires Hg(II) compounds as shown. Ethane–1,2–diol (b.p. 197°, high) and the 1,2–dithiol are especially useful in acetal and thioacetal formation (margin) because the second stage is favoured (entropy factor) when cyclic derivatives are formed.

Most imines[**2c(iv)**] are so easily hydrolysed that they cannot be isolated; some exceptions are in the margin. With carbonyl compounds having an αH secondary amines(R_2NH) give enamines, useful intermediates in synthetic work.

7.5 Reactions of enol forms and enolate anions

Many reactions of aldehydes and ketones are treated as part of 'aldol condensations' and are deferred until the next Section. Those in **7.5Re** are a separate group.

7.5Re

1 Halogenation at α positions

a as e.g. for discussion $R\text{-}CO\text{-}CH_3$ in H_2O + Hal_2 / HHal or KOH

$$R\text{-}CO\text{-}CH_3 \xrightarrow{\text{ i }} R\text{-}CO\text{-}CH_2Hal \xrightarrow{\text{ ii }} R\text{-}CO\text{-}CHHal_2 \xrightarrow{\text{ iii }} R\text{-}CO\text{-}CHal_3$$

R = e.g. Bu^t, Ph lacking αH

rate of each stage proportional to [ketone][HHal] or [ketone][KOH]

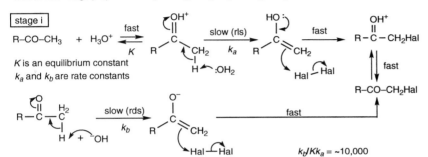

$\boxed{\text{stage i}}$

K is an equilibrium constant
k_a and k_b are rate constants

$k_b/Kk_a = \sim 10,000$

H_3O^+ **rate** = $Kk_a[\text{R–CO–CH}_3][\text{H}_3\text{O}^+]$ ‾OH **rate** = $k_b[\text{R–CO–CH}_3][\text{‾OH}]$

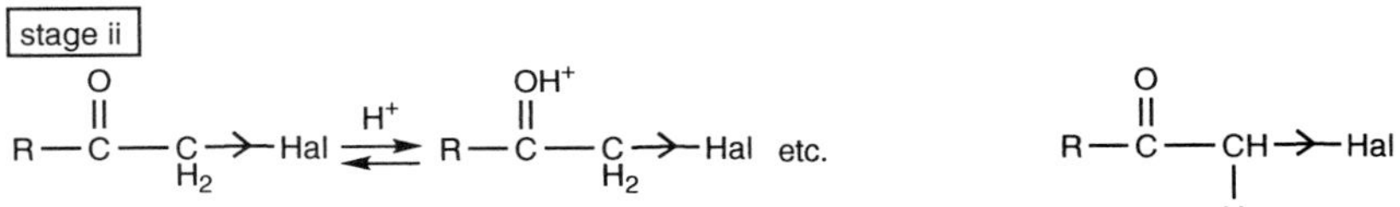

$\boxed{\text{stage ii}}$

Hal decreases tendency for H⁺ addn

acid catalysis: Kk_a decreases as go i to ii to iii

Hal increases acidity of --- ➤ H

base catalysis: k_b increases as go i to ii to iii

b e.g. $H_3C\text{-}C(\text{O})\text{-}CH_2\text{-}C\text{-}CH_3$ → more stable enol ... 70% ... 30% can stop at monobromo stage

$H_3C-C-C=CH_2$ less stable but formed faster than $H_3C-C=C-CH_3$

cannot stop at mono or dibromo stage

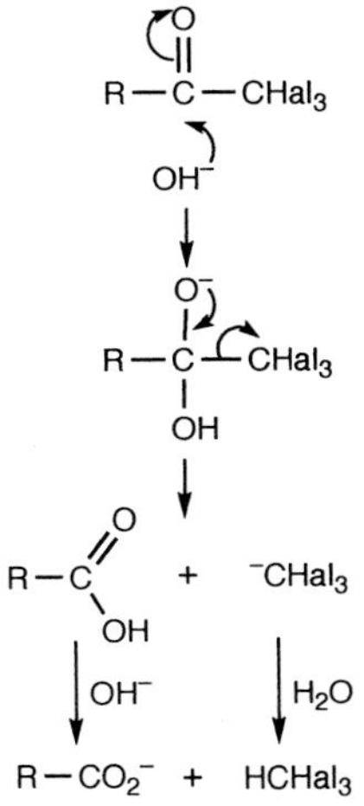

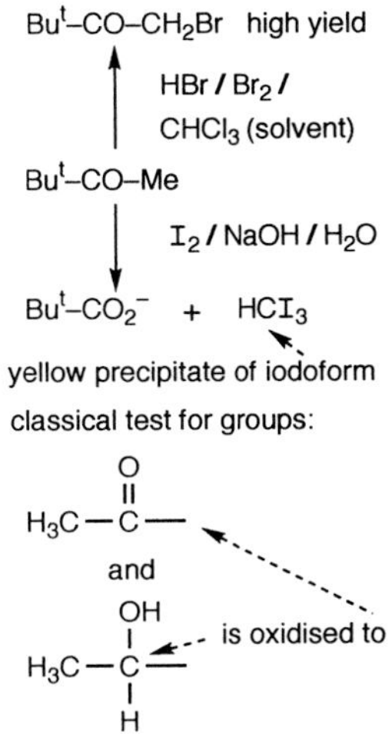

$Bu^t\text{-}CO\text{-}CH_2Br$ high yield

HBr / Br_2 / CHCl₃ (solvent)

$Bu^t\text{-}CO\text{-}Me$

I_2 / NaOH / H_2O

$Bu^t\text{-}CO_2^-$ + HCI_3

yellow precipitate of iodoform

classical test for groups:

$H_3C-C(\text{O})-$ and $H_3C-CH(OH)-$ is oxidised to

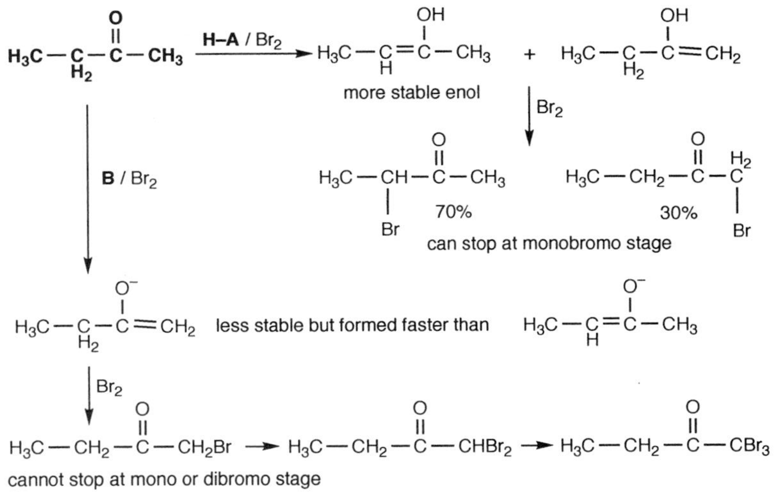

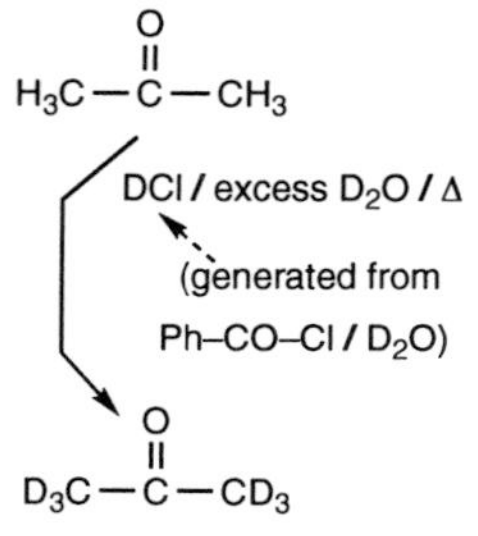

H3C—C(=O)—CH3

DCl / excess D2O / Δ

(generated from
Ph–CO–Cl / D2O)

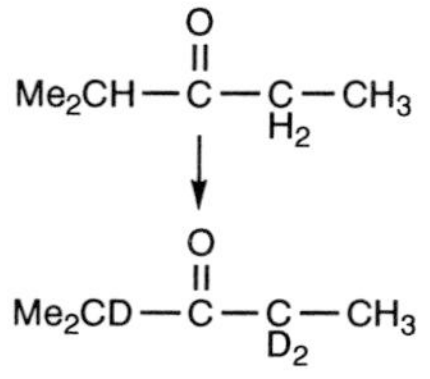

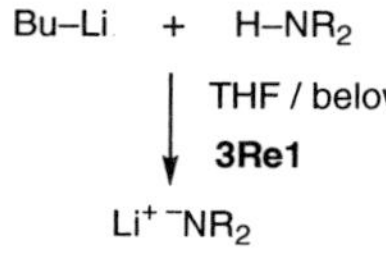

D3C—C(=O)—CD3

Me2CH—C(=O)—C(H2)—CH3

Me2CD—C(=O)—C(D2)—CH3

Bu–Li + H–NR2

THF / below 0°

3Re1

Li+ −NR2

very strong base, very poor nucleophile

(pK_a of H–NR2 ~38)

R = Me2CH

lithium diisopropylamide

abbreviated to LDA

R = Me3Si **7.4Re1f**
and others, R groups not
identical

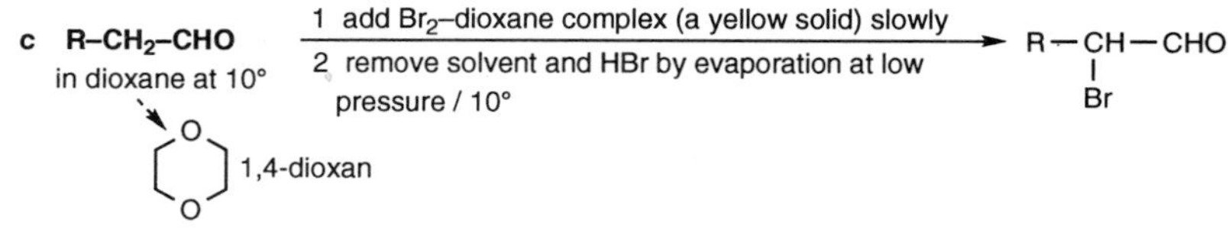

c R–CH2–CHO
in dioxane at 10°

1 add Br2–dioxane complex (a yellow solid) slowly
2 remove solvent and HBr by evaporation at low
   pressure / 10°

→ R—CH(Br)—CHO

1,4-dioxan

2 Deuteriation of ketones at α positions

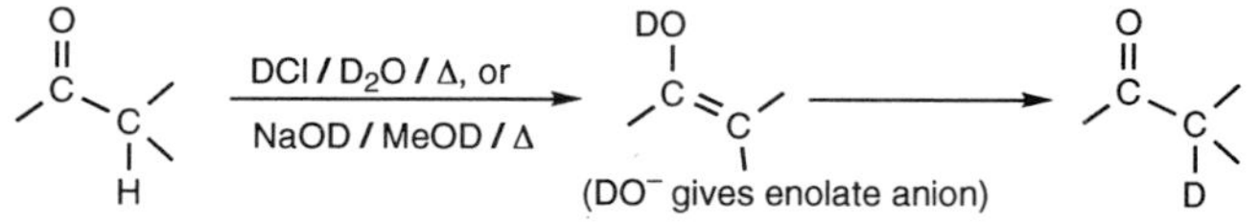

DCl / D2O / Δ, or
NaOD / MeOD / Δ

(DO⁻ gives enolate anion)

3 Alkylation

**a of ketones: direct,
using very strong base**

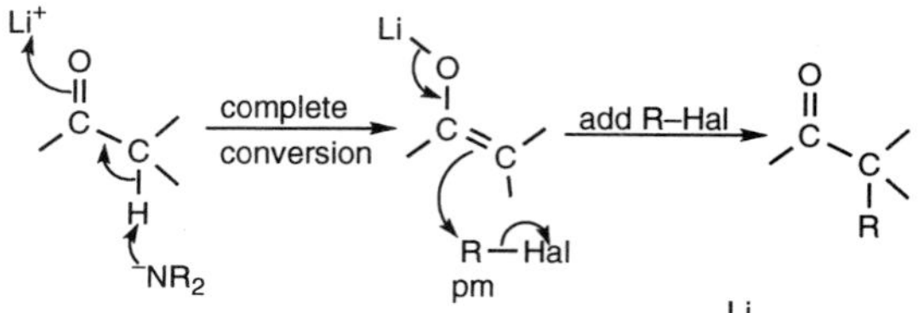

complete
conversion

add R–Hal

R—Hal
pm

e.g.

P

R–Hal

main product

add LDA / THF
slowly at 0°

P ←

at 0° with an excess of ketone present

add to LDA / THF
slowly at –20°

Q

R–Hal

main product

Q

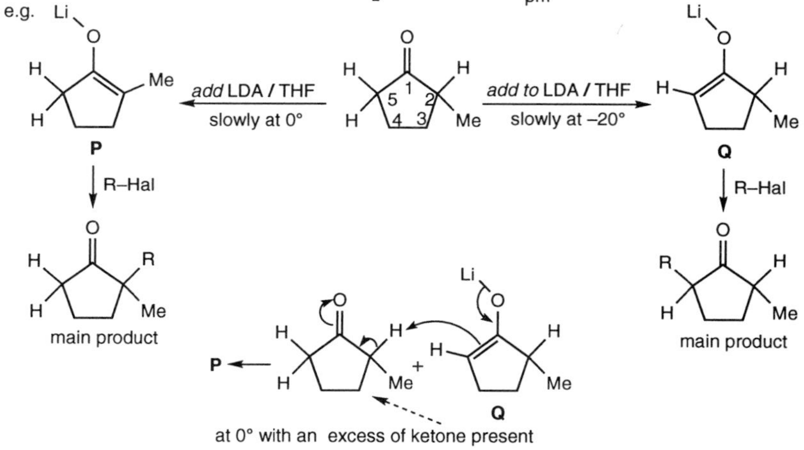

b of ketones: indirect, via 1,3-ketoesters

e.g.

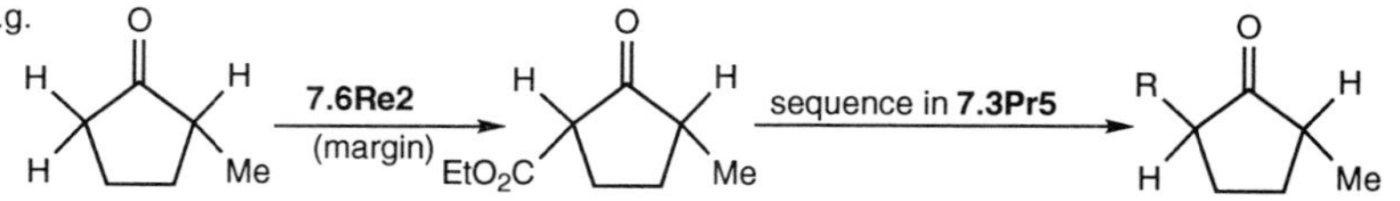

7.6Re2
(margin)

EtO2C

sequence in **7.3Pr5**

#**c of aldehydes: indirect via imines**

e.g.   Me2C—CHO
       |
       H

overall

Me2C—CHO
|
CH2Ph

7.4Re2c(iii)

H3O+

Et–MgBr
3Re1

PhCH2–Br

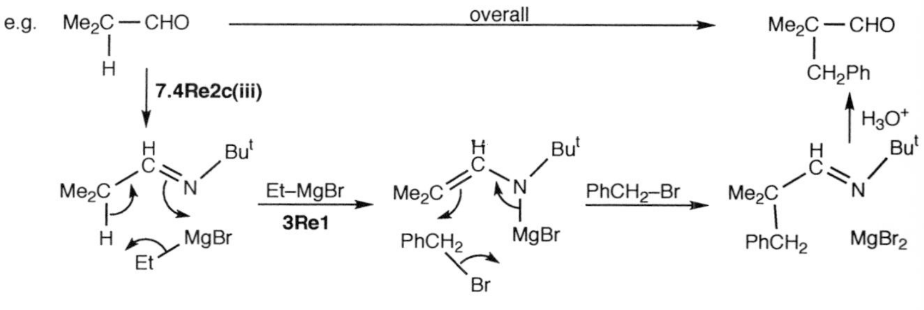

4 Enol esters

5 With methanal and amines 11Re6

Halogenation(**Re1**) is catalysed by acid of base; it is very slow in the absence of catalysts. The rates of the catalysed reactions are the same for chlorine, bromine and iodine. They and are not affected by the concentration of halogen but do depend on the concentration of acid or base. In the mechanisms shown the slow step is formation of enol or enolate anion. Subsequent reaction with halogen is fast. Steady–state treatment leads to useful expressions for the rates. ('Fast' and 'slow' are relative terms; on an absolute scale one reaction's slow stage might be faster than another's fast stage.) Under acid catalysis monohalogenation (stage i) is faster than di– or tri–halogenation (stages ii and iii). The presence of α halogen lowers the tendency of a carbonyl group to accept a proton (scheme); K, and hence the rate of stage ii, is decreased. Under base catalysis the stages become faster. The presence of α halogen makes the αH more acidic; k_b, and hence the rate, is increased. Monohalogenated ketones may be prepared under acid catalysis. The base catalysed reaction of a methyl ketone leads to the trihalogenated ketone, and unless the conditions are carefully controlled this is hydrolysed to the haloform (trihalomethane) and the salt of an acid. Thus from a methyl ketone(1mol), halogen(1mol) and an excess of alkali, unchanged ketone(0.66mol) and the salt of the acid(0.33mol) are formed. Aldehydes require milder conditions. The procedure shown gives satisfactory yields. Halogenocarbonyl compounds are lachrymators; the aldehydes are particularly unpleasant. In deuteriation(**Re2**) only αH (enolisable) is replaced by D.

Alkylation(**Re3**) by the older methods, for example K^+ $^-$OBu^t with R– Hal, is often indiscriminate because the ketone is only partially converted to its enolate. In modern procedures a very strong base is used at low temperature. Derivatives of lithium are especially useful(**3a**); these bulky bases (margin) are almost devoid of nucleophile character (see **1.12Ge**). The ketone is converted cleanly and completely to the enolate, and the other reactant is then added. In the example shown, 2–methylcyclopentanone, lithium enolates are depicted as covalent. This is merely to emphasise that they are more covalent than the fully ionic sodium and potassium derivatives. **Q** is formed under kinetic control; on steric grounds the C(5)H$_2$ is more accessible than the C(2)H. **P** has a more highly substituted double bond and is therefore the more stable enolate. This enolate is formed under thermodynamic control. Neither reaction is completely regiospecific. ($^#$Later courses, improvement by silylating the enolates.)

Re3b provides an indirect but very efficient alternative route to the 2– Me,5–R–cyclopentanone.$^#$**Re3c** illustrates an ingenious method for alkylating aldehydes; it has not yet been established as a general method. Enol–esters(**Re4**) are readily formed by both aldehydes and ketones.

7.6 Aldol and related condensations

Scheme **7.6Re** covers three condensations. [Condensation, a woolly term, is the joining together of two molecules by a process in which the constituents of a simple molecule (for example H$_2$O, EtOH) are lost.] The common theme is that *an aldehyde, ketone or ester acting as an electrophile* combines with *the anion of an aldehyde, ketone or ester acting as a nucleophile*. Although esters come later (Section 8.4) their use in these condensations is included here.

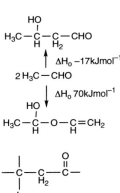

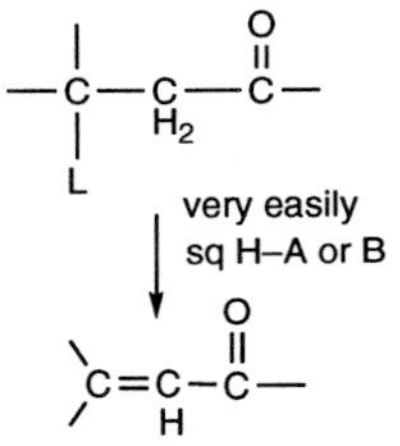

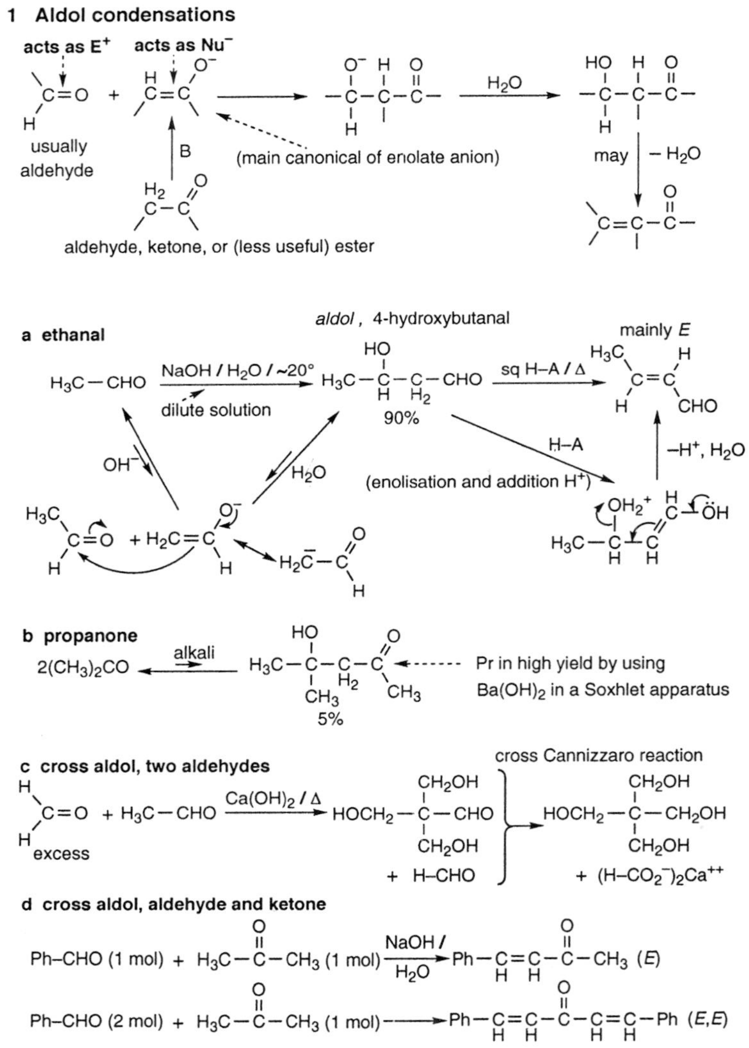

Cannizzaro reaction

only when no αH in R e.g. Ph

R–CHO / NaOH / H$_2$O / Δ

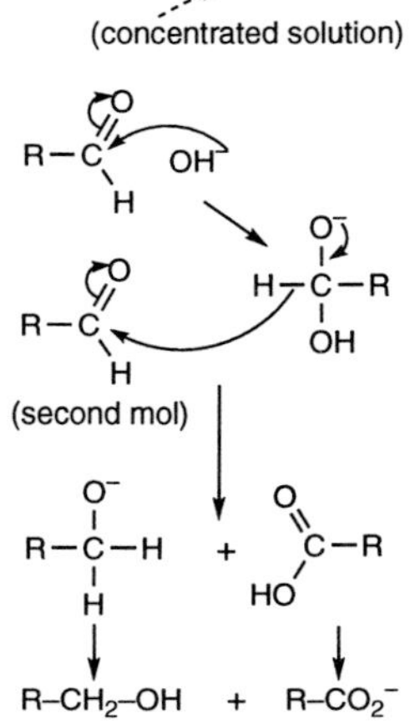

(concentrated solution)

7.6Re

1 Aldol condensations

acts as E$^+$ acts as Nu$^-$

(main canonical of enolate anion)

aldehyde, ketone, or (less useful) ester

usually aldehyde

a ethanal

aldol, 4-hydroxybutanal

H$_3$C–CHO $\xrightarrow{\text{NaOH / H}_2\text{O / ~20°}}$ H$_3$C–C–C–CHO $\xrightarrow{\text{sq H–A / Δ}}$ mainly *E*

dilute solution   90%

(enolisation and addition H$^+$)

b propanone

2(CH$_3$)$_2$CO $\xleftarrow{\text{alkali}}$ H$_3$C–C–C–C $\cdots\cdots$ Pr in high yield by using Ba(OH)$_2$ in a Soxhlet apparatus

5%

c cross aldol, two aldehydes

cross Cannizzaro reaction

C=O + H$_3$C–CHO $\xrightarrow{\text{Ca(OH)}_2 / \Delta}$ HOCH$_2$–C–CHO

excess

+ H–CHO

HOCH$_2$–C–CH$_2$OH

+ (H–CO$_2^-$)$_2$Ca^{++}

d cross aldol, aldehyde and ketone

Ph–CHO (1 mol) + H$_3$C–C–CH$_3$ (1 mol) $\xrightarrow{\text{NaOH / H}_2\text{O}}$ Ph–C=C–C–CH$_3$ (*E*)

Ph–CHO (2 mol) + H$_3$C–C–CH$_3$ (1 mol) $\longrightarrow$ Ph–C=C–C–C=C–Ph (*E,E*)

#e e.g. of **directed aldol condensation**

$Ph_2C{=}O$ + $H_3C{-}CHO$ →(overall)→ $Ph_2C{-}C{-}CHO$ (with HO and H_2)

cyclohexylamine / **7.4Re2c(iii)**

LDA → $H_2C{=}C{-}N$ (cyclohexyl), Li

$\xrightarrow[\text{at }-70°]{Ph_2CO}$ $H_2C{-}C{=}N$ (cyclohexyl)

$Ph_2C{-}$ OLi →(H_2O)→ (to product above)

2 Claisen condensations

acts as E⁺ **acts as Nu⁻**

ester + $C{=}C$ → $-C{-}C{-}C{-}R$ (or OR)

1,3-diketone or 1,3-ketoester

B ← ester or ketone

a ethyl ethanoate

$H_3C{-}CO_2Et$ $\xrightarrow{NaOEt / EtOH / \Delta}$ $H_3C{-}C{=}C{-}CO_2Et$ ($O^- Na^+$) $\xrightarrow{H{-}A}$ $H_3C{-}C{-}C{-}CO_2Et$ pK_a 11

$pK_a\ 24$ $pK_a\ 18$ EtO^-

Pr of acetoacetic ester

$H_3C{-}C$ (OEt) + $CH_2{=}C$ (O⁻)(OEt) ⇌ $H_2C{-}C$ (OEt)

$H_3C{-}C{-}C{-}CO_2Et$ (with EtO^-)

$H_3C{-}C{-}C{-}CO_2Et$ (OEt)

b ethyl 2-methylpropanoate

$Me_2CH{-}CO_2Et$ (1 mol) $\xrightarrow[\text{(1 mol)}]{\text{very strong base}}$ $Me_2\bar{C}{-}CO_2Et$ $\xrightarrow[\text{(second mol)}]{Me_2CH{-}CO_2Et}$ $Me_2CH{-}C{-}C{-}CO_2Et$ (with O, Me, Me)

c

$H_3C{-}C{-}CH_3$ (O) + $H_3C{-}CO_2Et$ with NaOEt / EtOH could give anions of 4 products:

figures are pK_a values

19 HO $(CH_3)_2C{-}C{-}C{-}CH_3$ 20 / 21

19 HO $(CH_3)_2C{-}C{-}C{-}CO_2Et$ 25

20 $H_3C{-}C{-}C{-}CO_2Et$ 11

product (high yield) is

20 9 20 $H_3C{-}C{-}C{-}C{-}CH_3$

acetylacetone
pentane-2,4-dione

$H{-}CO_2Me$ + $Ph{-}CH_2{-}CO_2Me$ E^+

$\downarrow$ NaOEt / EtOH / Δ

$Ph{-}CH{-}CO_2Me$ (with CHO)

(cyclohexanone) + CO_2Et / CO_2Et E^+

$\downarrow$ NaOEt / EtOH / Δ

(2-oxocyclohexanecarboxylate with CO_2Et)

(cyclopentanone, Me, positions 5 1 2 4 3) + $EtO{-}C{-}OEt$ E^+ diethyl carbonate

$\downarrow$ NaH

$EtO_2C{-}$ (cyclopentanone ring, Me, positions 5 1 2 4 3)

Re3

a

b

c

(?configuration)

3 Knoevenagel condensations

use in synthesis

aldehyde or ketone

pK_a ~10–13   electron withdrawing group $Nu^- = e.g.$ $^-CH(CO_2Et)_2$

most useful W = CO–R or CO_2Et, reacts with aldehydes and reactive ketones

W = CN *ethyl cyanoethanoate*, reacts with aldehydes and ketones

mechanism

In margin   **a** with **cyclohexanone**   **b** with **pentanal**   **c** with **Ph–CO–Me**

d Doebner modification

sq   as base /

pyridine as solvent / Δ   96%

All the steps of the aldol condensation(**Re1**) are reversible; indeed, under appropriate conditions an aldol, a 1,3–hydroxycarbonyl compound, reverts to starting materials. Thus, the condensations occur under thermodynamic control. Dilute alkali suffices as a base, and the reactions do not require a specific amount of alkali. Dehydration of an aldol is very easy because it produces a conjugated (stabilised) system. This dehydration is merely one example of a general tendency (margin). The parent reaction, with ethanal(**Re1a**), may be depicted according to taste as involving either canonical of the enolate anion; the crucial point is that the electrophilic component adds to the C of the anion. (See margin for enthalpy consideration.) There is a sharp contrast between the self–condensations of ethanal and propanone; the factors discussed in **7.4Re** are in play. A 'cross' condensation, different compounds as electrophiles and nucleophiles, is successful only when the reactants have different propensities. In **Re1c** methanal, a very reactive electrophile lacking αH, reacts with the anion of ethanal to give a trisubstituted ethanal which then undergoes a cross Cannizzaro reaction. (See margin for the standard Cannizzaro reaction. In the last stage of **1c** methanal is much more reactive than the trisubstituted aldehyde for OH^- addition.) In **1d** dehydration of the 1,3–hydroxyketone from benzaldehyde (no αH, a rather unreactive aldehyde) and propanone gives an extended conjugated system; the process is so easy that, in the presence of alkali, it cannot be prevented. At equilibrium the amount of hydroxyketone would be low, but the reaction is pulled through by the irreversible dehydration.[#]The overall outcome of **1e** is remarkable; in an alkaline mixture of ethanal and benzophenone, the ethanal would self–condense rather reacting with the unreactive ketone. Thus an 'unnatural' condensation has been realised.

The condensations in **Re2** (named after Claisen, 1851–1930, another illustrious organic chemist) are also reversible. Two points are to be stressed. Generally the species formed is *the anion of the product not the product itself.* The second is the corollary: *at least one mol of the base is required to form one mol of the anion.* Thus in **2a** the driving force is the formation of a stable anion. The product is the best known 1,3–ketoester, and is used for example as in **7.3Pr5**. Self–condensation of ethyl 2–methylpropanoate(**2b**) does not lead to a stable anion. Sodium ethoxide is therefore not effective; a very strong base is required, as shown. (There is a difficulty here. Some of the early very strong bases have been superseded, and to report their uses would be unprofitable. It is tempting to record, for example, LDA as the base in **2b**, but in fact the reaction has not been carried with LDA. In such cases the base is therefore not specified: a chemist wishing to repeat the reaction would choose a base by consulting modern procedures for similar reactions.) Of the four possible products in **2c** pentane–2,4–dione has the most acidic H; its anion is the one formed by thermodynamic control under basic conditions. Some successful cross Claised condensations are shown in the margin. They meet the requirement for differently disposed components.

It is convenient but not historically accurate to group all the material in **Re3** as Knoevenagel condensations (pronounced No–ven–ah–gel, with a hard 'g' as in 'get'). The reactions proceed well under mild conditions. They are therefore applicable to both aldehydes and ketones. Examples **3a,b** and **c** are in the margin. The products are αβ–unsaturated esters or ketones. In these the behaviour of the double bond is markedly different from that of an isolated double bond(**5Ge1**). The tendency for electrophilic addition is decreased but *nucleophilic addition occurs readily.* ([#]This is the Michael addition, a valuable means of building up more complex molecules.) **Re3d** is a convenient, direct and very efficient method for converting an aromatic aldehyde into an acid which has two more C atoms.

8 Carboxylic acids and derivatives

This chapter is divided into sections:

8.1   Trends in general reactions

8.1Ge

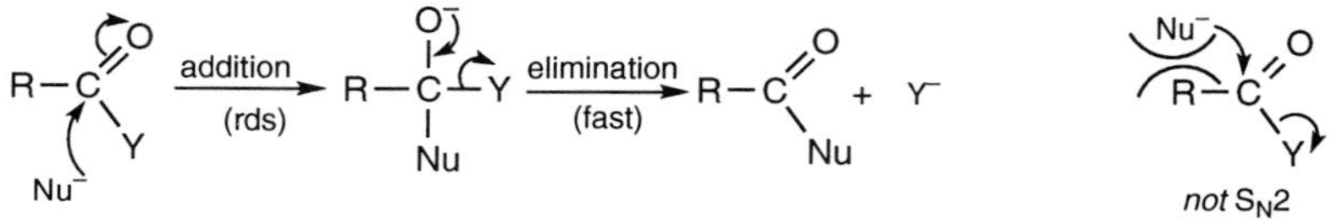

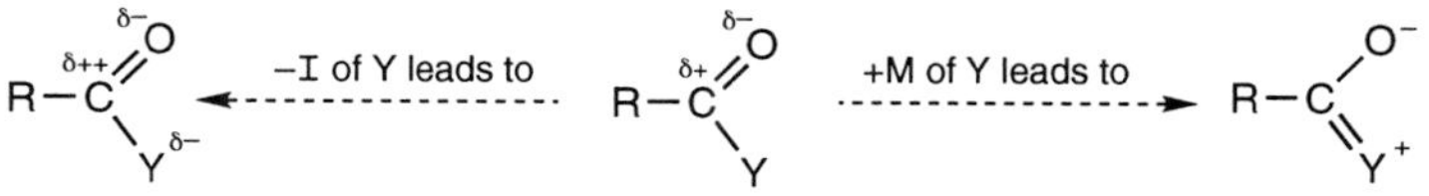

in **1,2,3** and **4** reactivity increases in direction of arrow

1   Reaction with Nu⁻

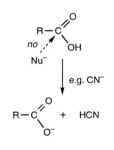

		aldehydes and		includes NHR^1 and NH_2
$Y = Cl$	$Y = O-CO-R$	$\begin{array}{c}\diagdown\\C=O\\\diagup\end{array}$	$Y = OR^1$	$Y = NR_2^1$
acyl chlorides	anhydrides	ketones	esters	amides
$-I \gg +M$	$-I > +M$	**7.1Ge**	$-I < +M$	$-I \ll +M$

2   Reduction with LiAlH₄   (in **2,3** and **4** products obtained after work-up)

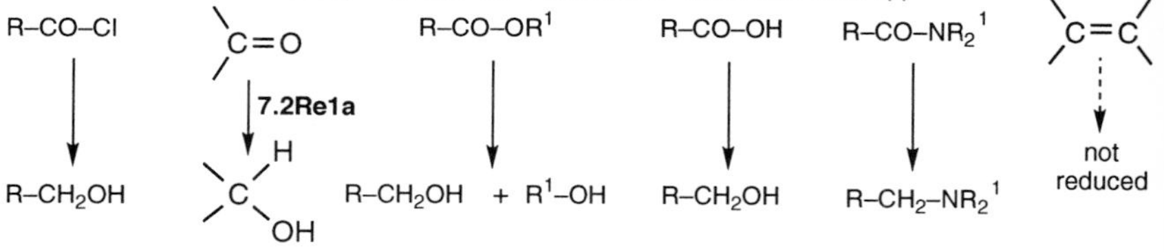

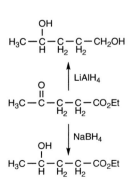

R–CO–Cl	$\begin{array}{c}\diagdown\\C=O\\\diagup\end{array}$	R–CO–OR¹	R–CO–OH	R–CO–NR₂¹	$\begin{array}{c}\diagdown\quad\diagup\\C=C\\\diagup\quad\diagdown\end{array}$
↓	↓ **7.2Re1a**	↓	↓	↓	↓
R–CH₂OH	$\begin{array}{c}\diagdown\;\;H\\C\\\diagup\;\;OH\end{array}$	R–CH₂OH + R¹–OH	R–CH₂OH	R–CH₂–NR₂¹	not reduced

3   Reduction with NaBH₄

R–CO–Cl	$\begin{array}{c}\diagdown\\C=O\\\diagup\end{array}$	R–CO–OR¹	R–CO–OH	R–CO–NR₂¹	$\begin{array}{c}\diagdown\quad\diagup\\C=C\\\diagup\quad\diagdown\end{array}$
↓	↓ **7.2Re1b**				
R–CH₂OH	$\begin{array}{c}\diagdown\;\;H\\C\\\diagup\;\;OH\end{array}$	not reduced			

4 Reduction with B₂H₆

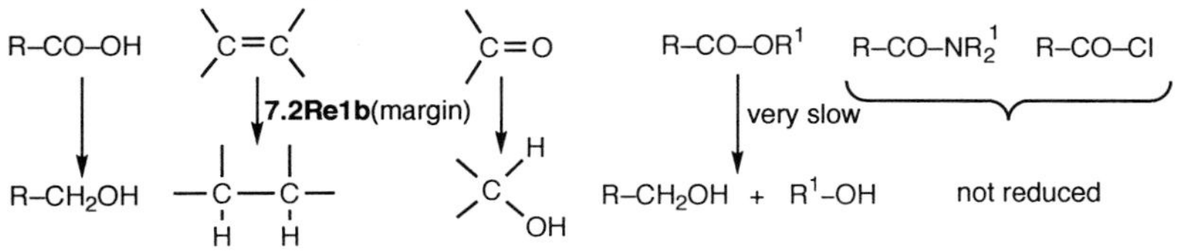

Mechanisms

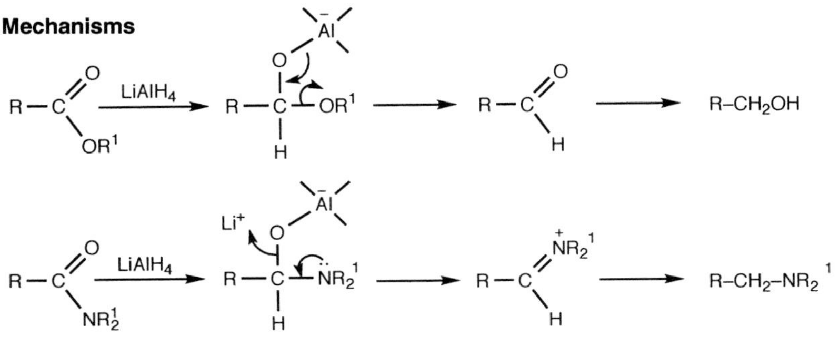

The chemistry of carboxylic acids and their derivatives is mostly concerned with *attack by nucleophiles* (**8.1Ge1**). The acids themselves do not react in this way because they alone have an acidic H. Even CN⁻, a very strong nucleophile (**1.11Ge**), acts as a base and removes the acidic H (margin). However after protonation the acids are prone to nucleophilic attack (**8.2Re3a**).

The general reaction shown at the top of **8.1Ge** consists of two steps, addition then elimination. A stereoelectronic factor favours approach of the nucleophile at 90° to the carbonyl group (**7.1Ge**) and this minimises steric repulsion. A possible alternative, direct S_N2 displacement of Y⁻ (scheme), is less favourable on both counts. The overall rate of the reaction is determined by the rate of the addition, the rate-determining step: the leaving group tendency of Y⁻ in the fast step is not important. Consideration of the inductive and mesomeric effects of Y in the series explains the relative reactivity (scheme). Acid chlorides, –I dominant, have a very positive carbonyl C and are the most reactive. Amides, +M dominant, have an approximately neutral carbonyl C and are the least reactive.

LiAlH₄(**Re2**) and NaBH₄(**Re3**) are *nucleophilic* reducing agents (**7.2Re1a,b**). LiAlH₄ is much the more powerful reagent. The order of reactivity and the identities of the groups which are not reduced are largely as expected. However there is an apparently surprising difference in the nature of the products from esters and amides with LiAlH₄. The crucial point (scheme) is that RO⁻ is a much better L⁻ than R₂N⁻(**1.11Ge**). Loss of some sort of Al–O group from the amides' intermediate is preferable to loss of R₂N⁻. This occurs in the fast step and, therefore, does not affect the rate. Selective reduction of the keto group in a keto–ester (margin) by NaBH₄ is a key stage in many syntheses.

Diborane(**Re4**) is an *electrophilic* reducing agent. So then, the order of reactivity is roughly the reverse of that in **1**, **2** and **3**. The remarkable feature is the very fast reduction of acids; regrettably, detailed mechanisms for diborane reductions have not been established.

8.2 Carboxylic acids

8.2 Pr

1 R–M 3Re2i

2 R–CH$_2$OH 7.2Re2a

3 RHC=O 7.2Re2b

4 R–CO–CH$_3$ 7.5Re1a(margin)

5 Nitriles 2.1Re1k

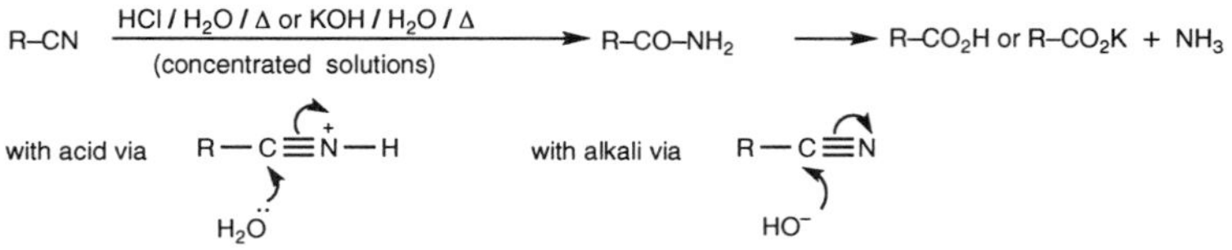

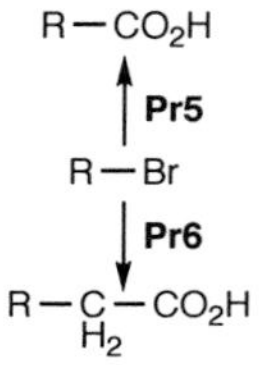

$$R–CN \xrightarrow[\text{(concentrated solutions)}]{\text{HCl / H}_2\text{O / }\Delta \text{ or KOH / H}_2\text{O / }\Delta} R–CO–NH_2 \longrightarrow R–CO_2H \text{ or } R–CO_2K + NH_3$$

with acid via $R–C\overset{+}{\equiv}N–H$ with alkali via $R–C\equiv N$

$H_2\overset{..}{O}$ HO^-

6 *Malonic ester*   Diethyl propanedioate

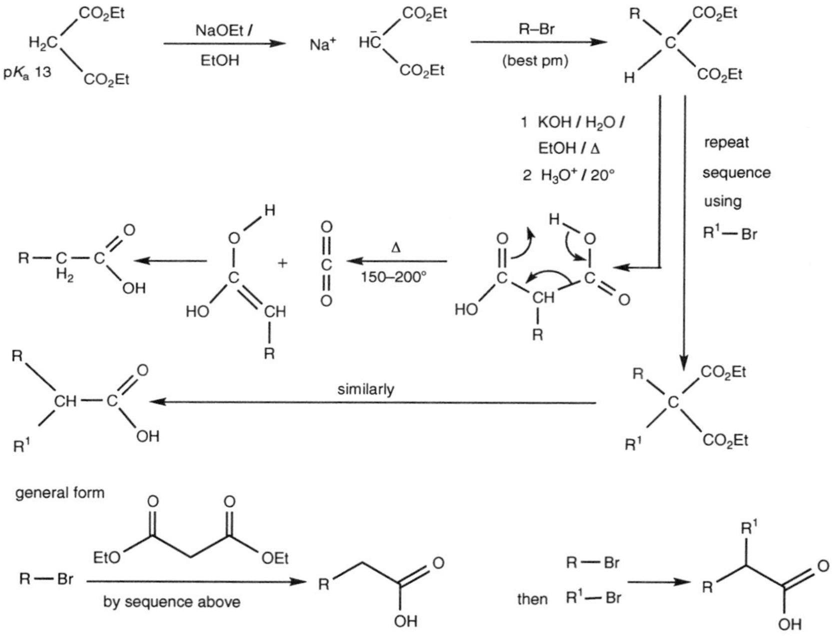

R–CO$_2$H

↑ **Pr5**

R–Br

↓ **Pr6**

R–C(H$_2$)–CO$_2$H

industrial process

Cl$_2$C=CHCl

↓ H$_2$SO$_4$(75%) / H$_2$O(25%) / 140°

Cl–C(H$_2$)–CO$_2$H

Re6

H$_3$Ċ–CO$_2$H

laboratory

↓ K$_2$CO$_3$

Cl–C(H$_2$)–CO$_2$K

↓ KCN

NC–C(H$_2$)–CO$_2$K

↓ HCl / EtOH / Δ

H$_2$C(CO$_2$Et)(CO$_2$Et)

$$H_2C(CO_2Et)_2 \xrightarrow[\text{EtOH}]{\text{NaOEt /}} Na^+ \; H\bar{C}(CO_2Et)_2 \xrightarrow[\text{(best pm)}]{\text{R–Br}} RCH(CO_2Et)_2$$

pK$_a$ 13

1 KOH / H$_2$O / EtOH / Δ
2 H$_3$O$^+$ / 20°

repeat sequence using R^1–Br

Δ 150–200°

general form

$$R–Br \xrightarrow[\text{by sequence above}]{EtO-CO-CH_2-CO-OEt} R–CH_2–CO_2H$$

$$\xrightarrow[\text{then } R^1–Br]{R–Br} R–CHR^1–CO_2H$$

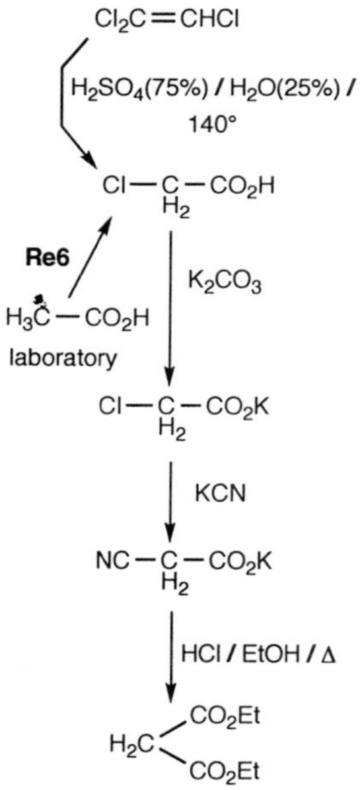

7 R–CO–Cl ⟶ **R–CH$_2$–CO$_2$H** 8.3Re3

Of the preparations not covered earlier **Pr5** and **6** are the important ones. Nitriles are readily obtained from alkyl halides; hydrolysis gives acids in high yield. The general sequence in **6,** based on malonic ester, recalls that in which ketones are prepared from acetoacetic ester(**7.3Pr5**).

The reactions of carboxylic acids are collected in **8.2Re.**

8.2Re

1 As acids

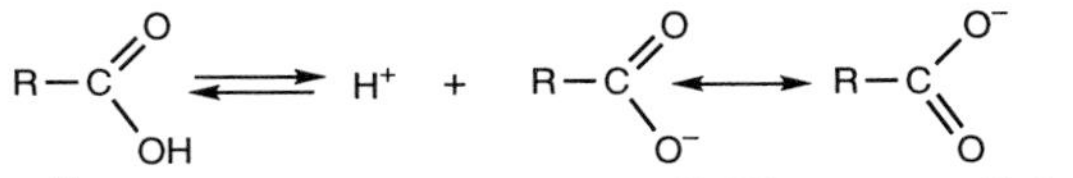

R =	H	Me	Et	Cl–CH$_2$	Cl$_3$C
pK_a	3.7	4.8	4.9	2.8	0.9

1.7Ge

2 Decarboxylation

no good one–stage method for simple acids; use e.g.   R–CO$_2$H $\xrightarrow{\textbf{2.1Pr4}}$ R–Br $\xrightarrow{\textbf{2.1Re4}}$ R–H

some types decarboxylate on heating:

W = electron withdrawing atom or group

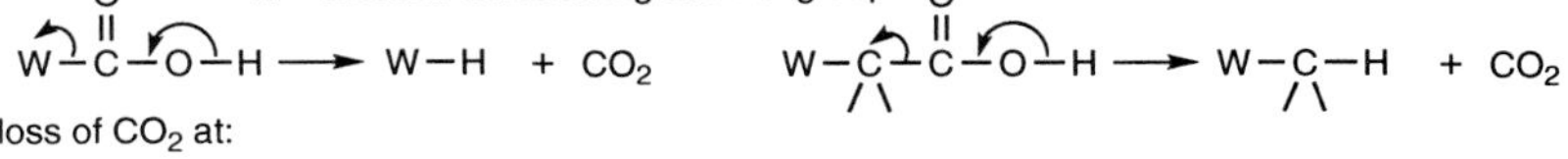

loss of CO$_2$ at:

20°   HO–CO–OH   Cl–CO–OH   H$_2$N–CO–OH   (these cannot be isolated)

20–50°   R–C–C–CO$_2$H (can be isolated at low temperature)   cyclic ts **7.3Pr5**

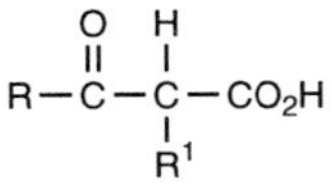

100–150°   Cl$_3$C–CO–OH   O$_2$N–CH$_2$–CO$_2$H   NC–CH$_2$–CO$_2$H

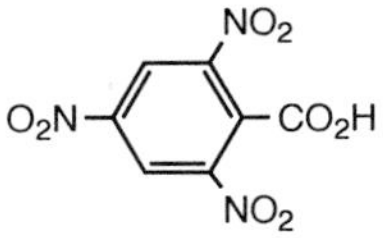

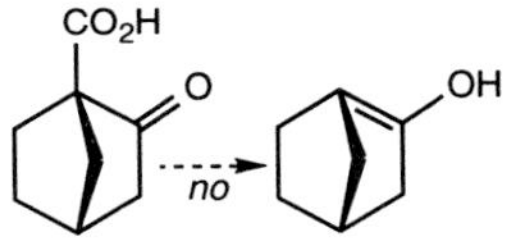

150–200°   HO$_2$C–C–CO$_2$H   cyclic ts **8.2Pr6**

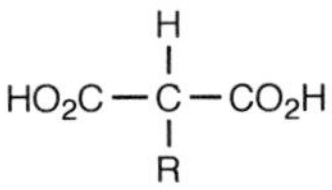

200–250°   3-enoic acids
             (βγ-unsaturated acids)

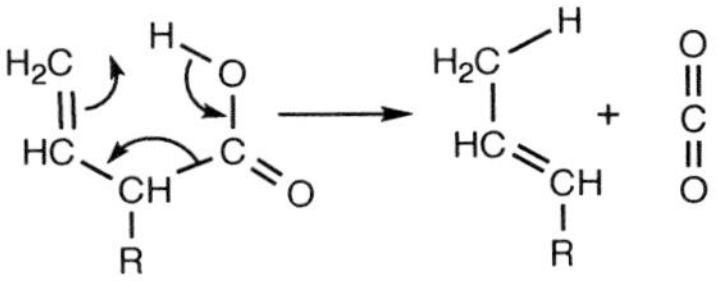

stable at 200°   impossible

3 Conversion to esters

a with R^1–OH   R–CO$_2$H + R^1–OH $\underset{\sim 5\% \text{ by weight of R}^1\text{OH}}{\overset{\text{H}_2\text{SO}_4 \text{ or HCl} / \Delta}{\rightleftharpoons}}$ R–CO$_2$R^1 + H$_2$O

(excess)

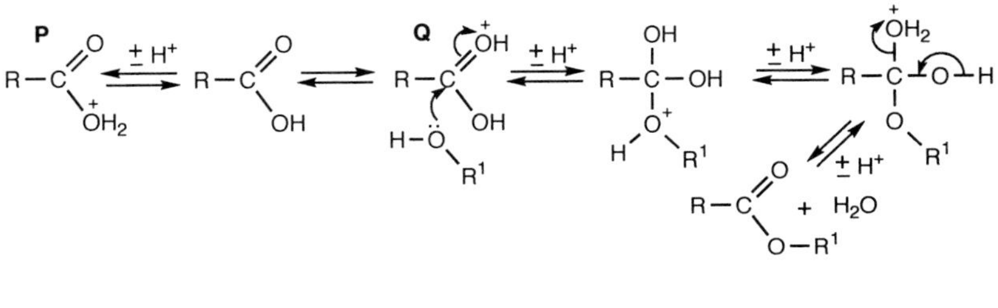

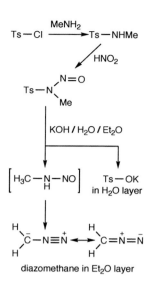

MeNH$_2$
Ts—Cl ⟶ Ts—NHMe

↓ HNO$_2$

Ts—N—N=O
       |
       Me

↓ KOH / H$_2$O / Et$_2$O

$\left[\text{H}_3\text{C}-\underset{\text{H}}{\text{N}}-\text{NO}\right]$ Ts—OK
                                   in H$_2$O layer

$\underset{H}{\overset{H}{C}}-\overset{+}{N}\equiv N \longleftrightarrow \underset{H}{\overset{H}{C}}=\overset{+}{N}=\bar{N}$

diazomethane in Et$_2$O layer

b with diazomethane

$H_2\bar{C}-\overset{+}{N}\equiv N$ ⟶

e.g. Me$_3$C—CO$_2$H ⇌ H$^+$ + Me$_3$C—C—O$^-$ H$_3$C—$\overset{+}{N}\equiv N$
                                              ‖
                                              O

H$_2$CN$_2$ / Et$_2$O /
20° / 10 mins ↓

(96%) Me$_3$C—CO$_2$Me + N$_2$ ⟵

c salts with alkyl halides
                                R^1–Br / Et$_2$O / Δ
R–CO–OAg ⟶ R–CO–OR^1 + AgBr   **2.1Re1c**
              best primary group

d with alkenes
e.g.   - - - b.p. –5°
R–CO–OH   sq H$_2$SO$_4$ / Me$_2$C=CH$_2$ / sealed apparatus / 0° / 24 hours ⟶ R–CO–O–CMe$_3$

                    H$^+$ ⟵ - - - - - - - - - - - - -
            $\overset{+}{C}$Me$_3$ O—CMe$_3$ O—CMe$_3$
R–C   ⟵   R–C   ⟶ R–C   + H$^+$
    ⟍O—H $\overset{+}{O}$–H   ‖
                                                           O

4 Pr of acyl chlorides

                              oxalyl chloride   ethanedioyl dichloride
                         O O ⟵ - - -
                         ‖ ‖
R–CO$_2$H Cl—C—C—Cl / C$_6$H$_6$ / Δ ⟶ R–CO–Cl + CO + CO$_2$ + HCl

R–CO$_2$Na ⟶ R–CO–Cl + CO + CO$_2$ + NaCl

SOCl$_2$ / Δ ↓   PCl$_5$ / C$_6$H$_6$ / warm
                    ⟶ R–CO–Cl + POCl$_3$ + HCl

R–CO–Cl + SO$_2$ + HCl

5 Pr of acid anhydrides
                                              R–CO–Cl / Δ
                  R–CO–O–CO–R + NaCl ⟵ R–CO$_2$Na
         R–CO–Cl /
R–CO$_2$H ⟶ R–CO–O–CO–R + HCl

       ⟨pyridine⟩ / Δ

6 Bromination (and chlorination) at αC   Hell–Volhard–Zelinsky

        O   sq P / Br$_2$ / Δ   O   OH
        ‖   ‖ |
R—C—C   ⟶   sq R—C—C   ⟷   R—C=C
    H$_2$ ⟍OH   H$_2$ ⟍Br   H   ⟍Br

no ↕   ↓ sq PBr$_5$ ↓ Br$_2$
        OH   H
        | | O
R—C=C   sq R—C—C
    H   ⟍OH   regenerates - - - | ⟍Br
no enol detected   and   Br
                H   O   produces   +
                | ‖   O
        R—C—C   more R—C—C
                | ⟍OH   H$_2$ ⟍OH
                Br

7 Conversion to amines   Schmidt

$$R\text{–}CO_2H \xrightarrow{\text{1 sq } H_2SO_4 / HN_3 \text{ (hydrazoic acid)} / CHCl_3 / \Delta \qquad 2\ H_2O} R\text{–}NH_2$$

↑ loss of CO_2

$R\text{–}NH\text{–}CO\text{–}OH$
carbamic acids

$R\text{–}C\equiv \overset{+}{O}$ ⎤ acylium ion

$R\text{–}\overset{+}{C}=O$ ⎦

$R\text{–}N=C=O$ + N_2
alkyl isocyanates

Methanoic acid *(formic acid)*, $H\text{–}CO_2H$, is unique among the acids in being a *reducing agent*. (An example of its use is in **11Pr4.**) The others are very resistant to oxidation. The acidity of $R\text{–}CO_2H$ stems from the stability of the derived anion(**8.2Re1**). A species whose canonicals are of equal energy, as here, may be regarded as having maximum resonance stabilisation. Methanoic acid is the strongest acid (scheme). Beyond propanoic *(propionic)* acid there is little change in pK_a. Substitution of Hal at the αC markedly increases acid strength. The trends can be rationalised in terms of the inductive effects of the groups attached to the carboxyl C but many other factors are involved.

Old direct methods for decarboxylating simple acids (e.g. heating with 'soda–lime') give very low yields and should be disregarded. The presence of an electron withdrawing group, attached to the CO_2H or the αC, facilitates decarboxylation(**Re2**). Cyclic transition states are favoured by some types of acid (scheme). For example simple 1,3–ketoacids lose CO_2 very easily. This depends on formation of an enol as the initial product: if the enol is geometrically impossible (margin) the acid is thermally stable.

Acid catalyzed esterification of acids(**Re3a**) is reversible. Formation of the ester is effected by using an excess of R^1OH (if cheap) or by removing the H_2O formed. An acid, acting as a very weak base, may be protonated at either oxygen to give the oxonium cations **P** and **Q**. Of these **Q** (two canonicals) is the more stable and is prone to nucleophilic attack. (Compare **8.1Ge.**) The intermediates en route to the ester have tetrahedral geometry: if R or R^1 is bulky there is repulsion between it and the other groups on the central C. Thus the esterification fails with tertiary R^1OH and with tertiary $R\text{–}CO_2H$.

Diazomethane(**Re3b**) reacts rapidly with all acids, even tertiary $R\text{–}CO_2H$, giving methyl esters in high yield. This reagent is a yellow poisonous gas (b.p.–24°). *Never attempt to isolate diazomethane: it is an explosive compound which has caused serious accidents.* A solution in ether, which is fairly safe, is used on a small scale in the fume cupboard. Protonation of diazomethane is followed by S_N2 reaction with the O^- of the carboxylate anion. The carbonyl C is not involved (compare **3a**). So although a bulky R inhibits attack at the carbonyl C it has little effect at the more distant O^-. Other diazoalkanes, $RHCN_2$, give the higher esters but these reagents are less readily prepared.

Although the Ag salt method(**3c**) is expensive it is useful when neutral conditions are required. Reaction **3d** involves a carbocation intermediate(Section 1.9); it is effective, therefore, only for preparing esters of tertiary R^1–OH.

Re4 and **5** are the standard preparations of acyl chlorides and acid anhydrides. If the acid contains a CC double bond it is necessary to avoid generating HCl. Oxalyl chloride with the Na salt of the acid is then the method of choice. Although **Re6** is not a very important reaction it has the distinction of a triple barrelled name. The reaction depends on a difference between acids and acyl halides with regard to enolisation. Enols of acids have not been detected: if they exist they are extremely unstable. Enolisation does occur with acyl halides and, as expected(**7.5Re1**), the enol is rapidly halogenated.

The Schmidt reaction(**Re7**) is restricted to acids in which the R group is stable to strong mineral acid. Isocyanates are intermediates here and in related reactions of esters and amides coming later. They add H$_2$O very easily to give carbamic acids, unstable compounds(**Re2**) which decompose to amines.

8.3   Acyl chlorides and acid anhydrides

The preparations are in **8.2Re4** and **5**. The chloride and anhydride of methanoic acid, H–CO–Cl and (H–CO)$_2$O, are unstable: they are not available as reagents in syntheses. Acyl chlorides are so reactive towards nucleophiles that the even more reactive bromides and iodides are rarely used. Acid anhydrides, somewhat less reactive, are more convenient to handle. 'Mixed' anhydides, R groups different, can be generated and used at temperatures below about 25°, but at higher temperatures they decompose to mixtures of the more stable simple anhydrides. A mixed anhydride (margin) serves as (H–CO)$_2$O in reactions requiring only low temperature.

8.3Re

1   With nitrogen nucleophiles

a   ammonia, pm and se amines

b   te amines

c   sodium azide

2 With oxygen nucleophiles

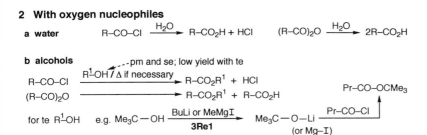

a water   R–CO–Cl $\xrightarrow{H_2O}$ R–CO$_2$H + HCl   (R–CO)$_2$O $\xrightarrow{H_2O}$ 2R–CO$_2$H

b alcohols

R–CO–Cl $\xrightarrow[\text{-pm and se; low yield with te}]{R^1\text{OH} / \Delta \text{ if necessary}}$ R–CO$_2$R^1 + HCl

(R–CO)$_2$O $\longrightarrow$ R–CO$_2$R^1 + R–CO$_2$H

for te R^{1}OH   e.g. Me$_3$C–OH $\xrightarrow[\textbf{3Re1}]{\text{BuLi or MeMgI}}$ Me$_3$C–O–Li $\xrightarrow{\text{Pr–CO–Cl}}$ Pr–CO–OCMe$_3$
(or Mg–I)

#3 With diazomethane

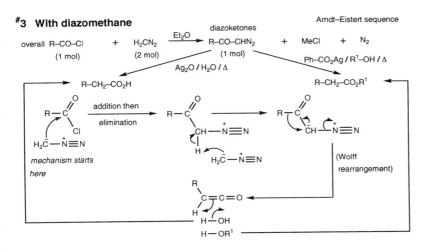

diazoketones   Arndt–Eistert sequence

overall R–CO–Cl + H$_2$CN$_2$ $\xrightarrow{Et_2O}$ R–CO–CHN$_2$ + MeCl + N$_2$
(1 mol)   (2 mol)   (1 mol)

$\xrightarrow{Ag_2O / H_2O / \Delta}$ R–CH$_2$–CO$_2$H $\xrightarrow{Ph-CO_2Ag / R^1-OH / \Delta}$ R–CH$_2$–CO$_2$R^1

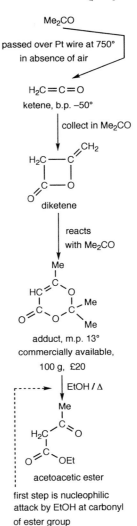

Me$_2$CO

passed over Pt wire at 750°
in absence of air

H$_2$C=C=O
ketene, b.p. –50°

collect in Me$_2$CO

diketene

reacts
with Me$_2$CO

adduct, m.p. 13°
commercially available,
100 g, £20

EtOH / Δ

acetoacetic ester

first step is nucleophilic
attack by EtOH at carbonyl
of ester group

Reactions(**8.3Re**) are shown for chlorides; in those specified the anhydrides may be used. To obtain the amide(**Re1a**) in high yield requires the presence of pyridine (scheme) or, in the alternative Schotten–Baumann procedure, NaOH. Without this extra base the yield is only 50% because the proton formed in the reaction removes R^1–NH$_2$ as its salt. Pyridine competes with R^1–NH$_2$ for this proton and, in effect, frees all the R^1–NH$_2$ for conversion to the amide.

With tertiary amines(**1b**) an amide cannot be formed. Here, a slower reaction (R$_3$N acting as a base) leads to ketenes, compounds having a CC double bond dittectly attached to a CO double bond. Only dialkyl ketenes (shown) can be isolated: those with one or two hydrogens form dimers. A product from the parent ketene (margin) is used to produce acetoacetic ester.

Reactions of chlorides and anhydrides with other nitrogen nucleophiles (e.g. hydrazine) have been omitted because the products are more conveniently obtained from esters(**8.4Re**). Oxygen nucleophiles(**Re2**) behave as expected. Esters of tertiary R^1–OH are prepared from the lithium alkoxides rather the alcohols.

#Reaction **3** is very useful for converting an acid to the next higher homologue in high yield (see scheme). The Ag$^+$ is not essential in all cases: some diazoketones can be rearranged photochemically or, in lower yield, by heat alone.

8.4 Esters

Scheme **8.4Pr** gives references to the preparations; **8.4Re** shows the reactions.

8.4Pr

1 $R-CO_2H$ 8.2Re3

2 $R-CO-Cl$, $(R-CO)_2O$ 8.3Re2b,3

3 R_2CO 7.4Re1e

8.4Re

1 Hydrolysis

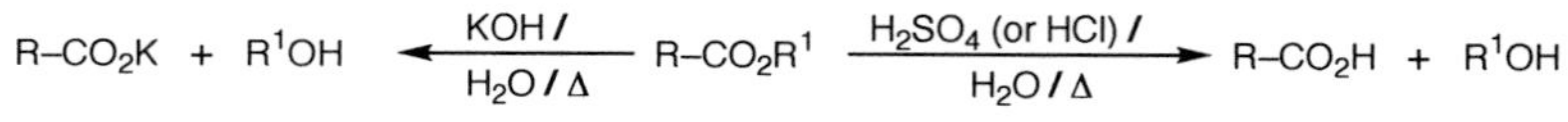

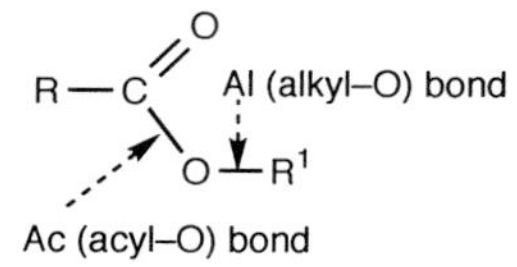

$$R-CO_2K + R^1OH \xleftarrow[H_2O\,/\,\Delta]{KOH\,/\,} R-CO_2R^1 \xrightarrow[H_2O\,/\,\Delta]{H_2SO_4\ (or\ HCl)\,/\,} R-CO_2H + R^1OH$$

Terminology:
A = slow step (rds or rls) involves ester
B = slow step involves conjugate acid of ester
subscript Ac or Al, which bond is broken
1 or 2, slow step is uni– or bi–molecular

Examples to illustrate commonest mechanisms:

a methyl benzoate with alkali $B_{Ac}2$

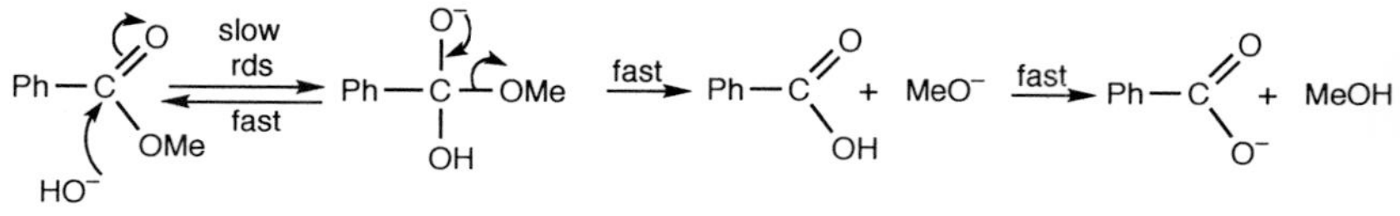

b ethyl butanoate with acid $A_{Ac}2$

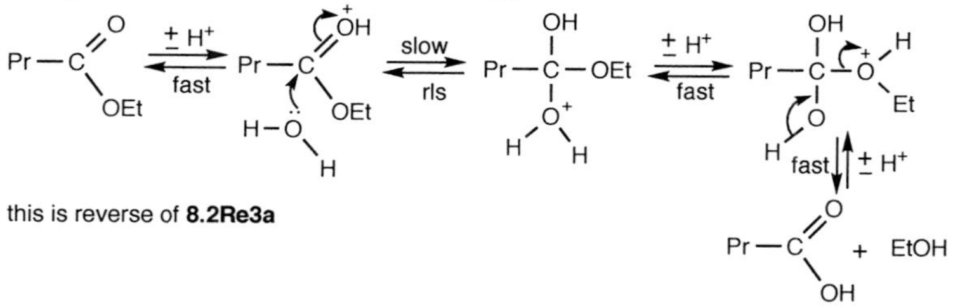

this is reverse of **8.2Re3a**

c t-butyl butanoate with acid $A_{Al}1$

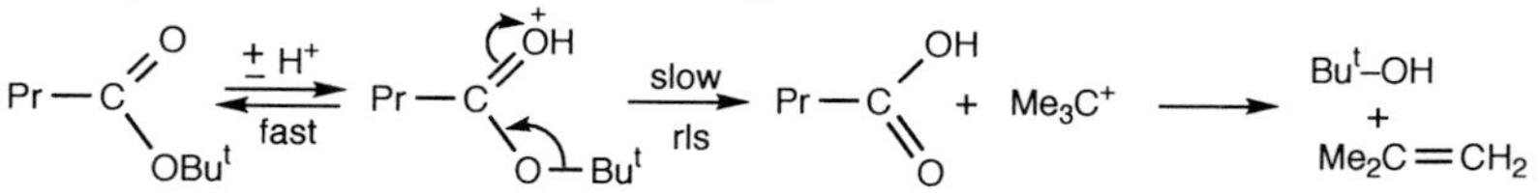

2 With nitrogen nucleophiles

a ammonia, pm and se amines

not generally useful

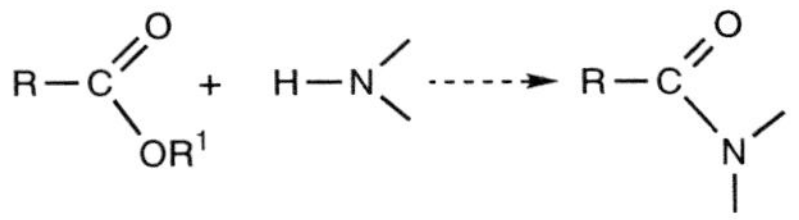

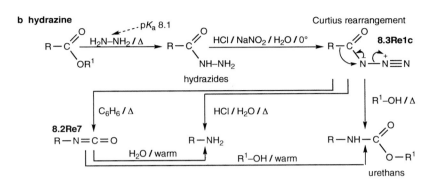

b hydrazine

hydrazides

urethans

3 With base in Claisen condensation 7.6Re2

See margin for intramolecular version, the Dieckmann reaction

#4 With Na, the acyloin reaction

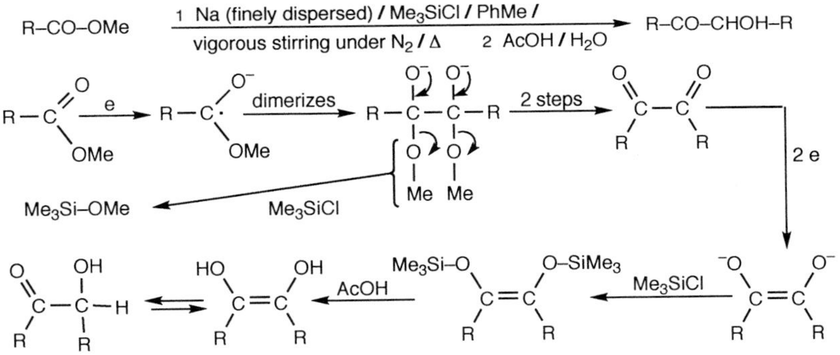

acyloins (α-hydroxyketones)

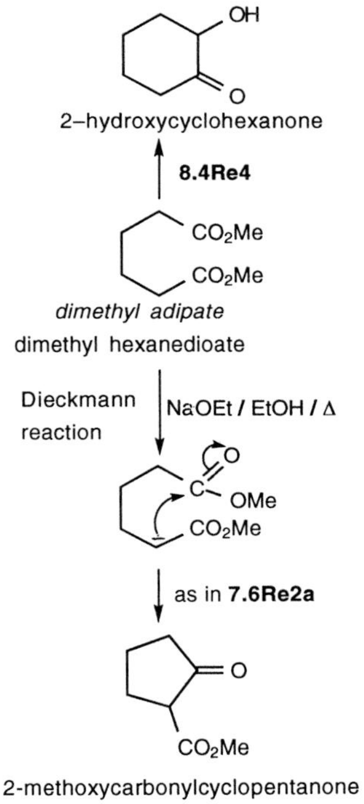

2–hydroxycyclohexanone

8.4Re4

dimethyl adipate
dimethyl hexanedioate

Dieckmann reaction | NaOEt / EtOH / Δ

as in **7.6Re2a**

2-methoxycarbonylcyclopentanone

The mechanism of hydrolysis(**8.4Re1**) has been studied intensively; the results are expressed in a special terminology. The terms A and B are generally stated to mean acid– and base–catalysed hydrolysis. This is not strictly true. The correct attributions of A and B, and the meanings of the other terms are summarised in the scheme. Of the eight possibilities (2 x 2 x 2) two, $A_{Al}2$ and $B_{Ac}1$, are unknown. Three common mechanisms are illustrated.

Reaction **Re2a**, generally slow, is a poor method for preparing amides. Compared with amines, hydrazine(**2b**) is weaker as a base but stronger as a nucleophile (reason unknown). The sequence ester to hydrazide to acyl azide and Curtius rearrangement gives primary amines in high yield.

The trimethylsilyl chloride in **Re4** traps the product as a silyl ether. It also removes the MeO⁻ which, if left free, leads to undesired by-products. Lack of space has precluded much about bifunctional compounds in this book. However **Re3** and **Re4** with the esters of dibasic acids are so important in carbocyclic chemistry that examples are included (margin). **Re3** is limited to the formation of five– and six–membered rings, but **Re4** (the acyloin reaction) gives high yields of rings ranging from small (four–membered) to large (twenty–membered).

8.5 Amides

8.5Pr

1 R–CO–Cl, (R–CO)₂O **8.3Re1a**

2 RCN

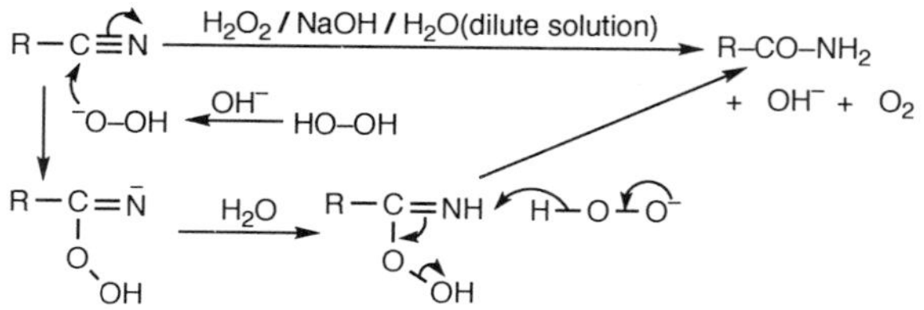

R—C≡N $\xrightarrow{\text{H}_2\text{O}_2\,/\,\text{NaOH}\,/\,\text{H}_2\text{O(dilute solution)}}$ R–CO–NH₂

⁻O–OH $\xleftarrow{\text{OH}^-}$ HO–OH

+ OH⁻ + O₂

R—C=N̄ $\xrightarrow{\text{H}_2\text{O}}$ R—C=NH
| |
O   O
  \OH   \OH

3 Oximes

Beckmann rearrangement

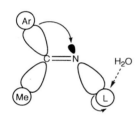

5Pr2b

Ar
  \
   C=O →
  /
Me

7.4Re2c(i)

Ar
  \
   C=N
  /   \
Me   OH

(*E*)

$\xrightarrow{\text{1 polyphosphoric acid }/\,\Delta}$ $\xrightarrow{\text{2 add ice}}$

Ar = Br—⟨ ⟩— (4-bromophenyl)

Ar
  \
   C=N
  /   \
Me   OH₂⁺

→

Ar
|
N
‖
C⁺
|
Me

↔

Ar
|
N⁺
|||
C
|
Me   HO—H

→

        O
       ‖
Me—C
       \
        NHAr

Me—C
  / \
 OH
  \
   NAr

enol of amide

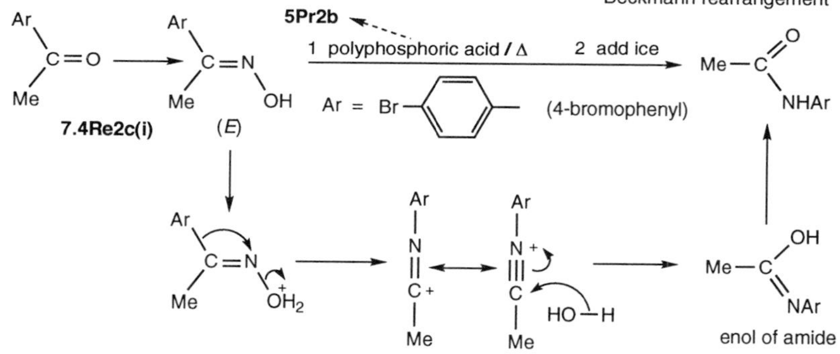

Reaction **8.5Pr1** is the most important preparative method. In the hydrolysis of nitriles(**8.2Pr5**), which requires vigorous conditions, amides are formed but hydrolyzed further to acids. Reaction **8.5Pr2**, designed to stop at the amide stage, is based on an unexpected feature: although H₂O₂ (pK_a12) is a stronger acid than H₂O (pK_a15.7) the anion ⁻O–OH is a stronger nucleophile than OH⁻. The stereochemistry of **Re3** is discussed a little later. The example shown is unusually 'clean'. Formation of the oxime gives almost entirely the (*E*) diasteroisomer. Rearrangement affords one amide in high yield.

8.5Re

1 As very weak acid and very weak base

$$R-C\begin{smallmatrix}O\\ \\NH\end{smallmatrix} \longleftrightarrow R-C\begin{smallmatrix}O^-\\ \\NH\end{smallmatrix} \overset{OH^-}{\rightleftharpoons} R-C\begin{smallmatrix}O\\ \\NH_2\end{smallmatrix} \longleftrightarrow R-C\begin{smallmatrix}O^-\\ \\\overset{+}{N}H_2\end{smallmatrix} \overset{H^+}{\rightleftharpoons} R-C\begin{smallmatrix}OH\\ \\\overset{+}{N}H_2\end{smallmatrix}$$

anion   **P** $pK_a\sim17$ **Q**   conjugate acid $pK_a\sim 0.5$

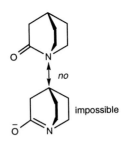

no

impossible

behaves as a ketone
and an amine

2 Hydrolysis and dehydration

$$R-CO-N\diagdown \xrightarrow[\textbf{8.2Pr5}]{} \begin{smallmatrix}R-CO-OH\\ \overset{+}{H-N}\diagdown\end{smallmatrix}$$

$$R-CO-NH_2 \xrightarrow[\text{(CF}_3-\text{CO)}_2\text{O / pyridine / }\Delta]{PCl_5 / \Delta \text{ or}} R-C\equiv N$$

$$R-CO-NHR^1 \xrightarrow{PCl_5 / \Delta} \begin{smallmatrix}R\\ \diagdown\\ C=N\\ \diagup \quad \diagdown\\ Cl \qquad R^1\end{smallmatrix}$$

chloroimines

3 Bromine (or chlorine) and alkali

Hofmann

$$R-CO-NH_2 \xrightarrow{Br_2 / NaOH / H_2O / \Delta} R-NH_2 + CO_2$$

$$\overset{\displaystyle\diagup\diagup OH^-}{}$$

$$R-CO-NH^- \xrightarrow{Br_2} R-CO-NH-Br \xrightarrow[\text{}]{OH^-} \rightleftharpoons R-C\begin{smallmatrix}O\\ \\N-Br\end{smallmatrix} \longrightarrow R-N=C=O + Br^-$$

$\xrightarrow{H_2O}$

8.2Re7 and 8.4Re2b

Amides are neutral compounds with only feeble acidic and basic properties(**8.5Re1**). They do not behave as ketones or as amines. **P** and **Q** are the canonicals. **Q**, the major contributor, lacks the partial positive on carbon characteristic of ketones and the lone pair on nitrogen characteristic of amines. The point is nicely illustrated (margin) by an amide in which the +M effect is inhibited.

There is a close resemblance between rearrangements **8.2Re7**(Schmidt), **8.4Re2b**(Curtius) and **8.5Re3**(Hofmann). Two more, **8.3Re3**(Wolff) and **8.5Pr3**(Beckmann), are of the same type. They share a common feature: one group migrates as an incipient carbanion and displaces a second (as a leaving group) which is in a trans orientation. This is illustrated (margin) for the Beckmann rearrangement.The basis is the general stereoelectronic effect of **5Ge2**. *In the transition state the groups are in an antiperiplanar arrangement.*

9 Alcohols   10 Ethers

The general tendencies are in **9Ge**. Most of the preparations and reactions have been encountered previously: references are in schemes **9Pr** and **9Re**.

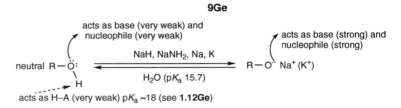

9Ge

acts as base (very weak) and nucleophile (very weak)

acts as base (strong) and nucleophile (strong)

NaH, NaNH$_2$, Na, K

neutral R—Ö:

$\ce{R-O^-}$ Na$^+$ (K$^+$)

H$_2$O (pK_a 15.7)

H

acts as H–A (very weak) pK_a ~18 (see **1.12Ge**)

9Pr

1 R–Hal 2.1Re1a

2 R–M 3Re1b,2a,2b,2c

3 Reduction of R(H)–CO–R 7.2Re1a,1b,1c **R–CO–Cl** 8.1Ge2,3

   R–CO$_2$H 8.1Ge2,4

4 $\ce{C=C}$ 5Re1b,3a,3b

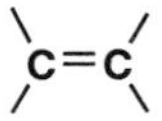

9Re

1 With R–M 3Re1

2 Oxidation 7.2Re2a,2b,3,4,5,6

3 With R–CO$_2$H 8.2Re3a **R–CO–Cl** 8.2Re4 **(R–CO)$_2$O** 8.2Re5

4 Pr of R–O–R^1 10Pr1

5 Oxidation of 1,2-diols

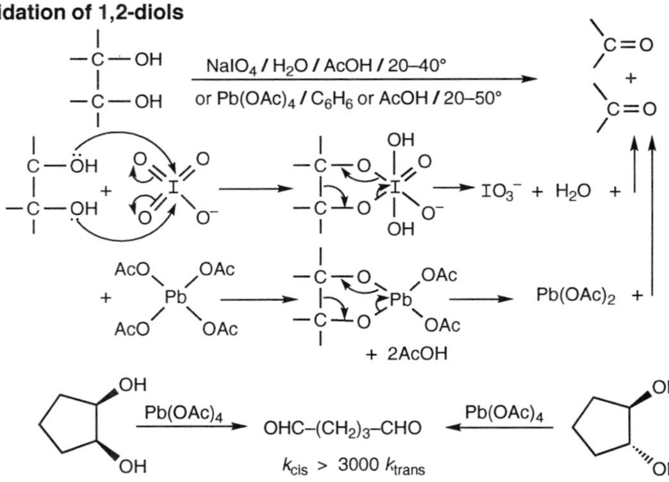

NaIO$_4$ and Pb(OAc)$_4$ oxidations:

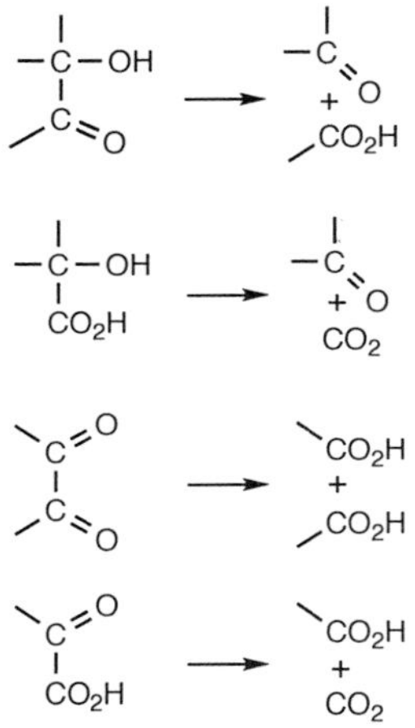

$-\underset{|}{\overset{|}{C}}-OH$ $-\underset{\diagdown}{\overset{|}{C}}{=}O$

$\underset{\diagup}{\overset{\diagdown}{C}}{=}O$ $+ \overset{O}{\underset{CO_2H}{}}$

$-\underset{|}{\overset{|}{C}}-OH$ $-\underset{\diagdown}{\overset{|}{C}}{=}O$

CO_2H $+ \overset{O}{\underset{CO_2}{}}$

$\underset{\diagup}{\overset{\diagdown}{C}}{=}O$ $\overset{\diagdown}{CO_2H}$

$+$

$\underset{\diagup}{\overset{\diagdown}{C}}{=}O$ $\underset{\diagup}{CO_2H}$

$\underset{CO_2H}{\overset{\diagdown}{C}}{=}O$ $\overset{\diagdown}{CO_2H}$

$+$

CO_2

NaIO$_4$ / H$_2$O / AcOH / 20–40°
or Pb(OAc)$_4$ / C$_6$H$_6$ or AcOH / 20–50°

$\ce{C=O}$ + $\ce{C=O}$

IO$_3^-$ + H$_2$O

AcO, OAc   Pb   AcO, OAc

Pb(OAc)$_2$ +

+ 2AcOH

OHC–(CH$_2$)$_3$–CHO

k_{cis} > 3000 k_{trans}

Pb(OAc)$_4$ Pb(OAc)$_4$

Reaction **9Re5** is concerned with sodium periodate and lead tetraacetate,very useful reagents which have similar oxidising properties. To dissolve $NaIO_4$, a salt, an aqueous medium is required. $Pb(OAc)_4$, a covalent compound, is soluble in some organic solvents and is decomposed by water. Thus the choice of reagent is determined by the solubility of the substrate. With both reagents a cyclic intermediate is formed in the first stage. This explains the difference in rates (margin) between cis cyclopentane–1,2–diol (easy ring formation) and the trans isomer (ring formation very difficult). The reagents cause fission of several other systems having 1,2–oxygen–containing groups (margin). $NaIO_4$ is involved in the method(**5Re3d**) for cleaving CC double bonds.

The chemistry of ethers is in schemes **10Ge, Pr** and **Re**

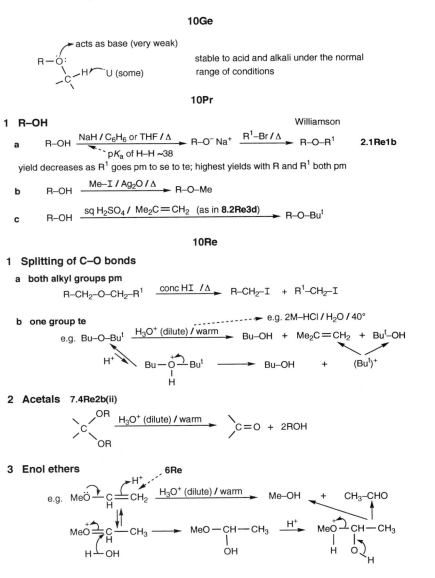

10Ge

acts as base (very weak)

$R-\ddot{O}:$

$\overset{|}{\underset{|}{C}}-H$ U (some)

stable to acid and alkali under the normal range of conditions

10Pr

1 R–OH

                                                        Williamson

a $R-OH \xrightarrow[\text{p}K_a \text{ of H–H} \sim 38]{\text{NaH / } C_6H_6 \text{ or THF / } \Delta} R-O^- Na^+ \xrightarrow{R^1-Br / \Delta} R-O-R^1$ **2.1Re1b**

yield decreases as R^1 goes pm to se to te; highest yields with R and R^1 both pm

b $R-OH \xrightarrow{Me-I / Ag_2O / \Delta} R-O-Me$

c $R-OH \xrightarrow{\text{sq } H_2SO_4 / Me_2C=CH_2 \text{ (as in } \mathbf{8.2Re3d})} R-O-Bu^t$

$Me-I + KO-Bu^t$

$\downarrow S_N2$

$Me-O-Bu^t$

$Bu^t-Br + KO-Me$

$\downarrow E1$

$Me_2C=CH_2$

10Re

1 Splitting of C–O bonds

  a both alkyl groups pm

         $R-CH_2-O-CH_2-R^1 \xrightarrow{\text{conc HI } / \Delta} R-CH_2-I + R^1-CH_2-I$

  b one group te

                             - - - - → e.g. 2M–HCl / H_2O / 40°

         e.g. $Bu-O-Bu^t \xrightarrow{H_3O^+ \text{ (dilute) / warm}} Bu-OH + Me_2C=CH_2 + Bu^t-OH$

                $H^+ \searrow$ $Bu-\overset{+}{\underset{\underset{H}{|}}{O}}-Bu^t \longrightarrow Bu-OH + (Bu^t)^+$

2 Acetals 7.4Re2b(ii)

         $\overset{\diagdown}{\underset{\diagup}{C}}\overset{OR}{\underset{OR}{\big\langle}} \xrightarrow{H_3O^+ \text{ (dilute) / warm}} \overset{\diagdown}{\underset{\diagup}{C}}=O + 2ROH$

3 Enol ethers   **6Re**

   e.g. $Me\ddot{O}-\overset{\underset{H}{|}}{C}=CH_2 \xrightarrow{H_3O^+ \text{ (dilute) / warm}} Me-OH + CH_3-CHO$

         $Me\overset{+}{O}=\overset{\underset{H}{|}}{C}-CH_3 \longrightarrow MeO-CH-CH_3 \xrightarrow{H^+} Me\overset{+}{O}-CH-CH_3$

           $H-OH$ OH H O

4 Oxirans *(epoxides)* 5Re3c

e.g. $Me_2C{-}CH_2$ (oxiran) $\xrightarrow[\text{or NaOH / H}_2\text{O / warm (slower reaction)}]{\text{H}_3\text{O}^+ \text{ (dilute) / warm}}$ $Me_2C{-}\underset{HO}{\overset{OH}{C}}H_2$

Ethers(**10Ge**) are ideal solvents for many reactions but only special types of ethers can be regarded as reagents. Atmospheric O_2 reacts slowly with ethers by a radical mechanism to give highly explosive peroxides. This applies particularly to ethers containing secondary R groups. Peroxides must be removed from ethers before they are heated; distillation of 'old' $(Me_2HC)_2O$, *(diisopropyl ether)*, has resulted in serious accidents.

The reaction used most frequently for preparing ethers is **10Pr1a**. Dependence of the yield on the structure of R and R^1 is as expected from the discussion of **2.1Ge2a**. Two pairs of starting materials which could, formally, lead to 2–methoxy–2,2–dimethylethane are shown in the margin; the yield is high with the first pair but the main product from the second pair is the alkene. Method **1b** has been widely used in carbohydrate chemistry for methylating primary and secondary alcohols. Preparation **1c** is restricted to ethers containing a tertiary R group.

Splitting simple ethers, two primary alkyl groups, requires vigorous treatment with the strongly nucleophilic hydroiodic acid (**10Re1a**). Four special types are shown in **Re1b, 2, 3** and **4**. These undergo useful transformations under *mild acid conditions*. The first three are stable to alkali. Ring opening of the fourth, the oxirans(**Re4**), occurs with acid and alkali; the same diol is formed but by different mechanisms (margin). Nucleophilic attack occurs at the C which is less hindered and less negative. However, as expected from the earlier discussion(**5Re2**), the alternative ring opening occurs with the protonated oxiran.

O = ^{18}O

$Me_2\underset{HO}{\overset{}{C}}{-}\overset{OH}{C}H_2$

$\uparrow$ NaOH / H_2O

$\overset{Me}{\underset{Me}{}}C\overset{OH}{-}CH_2$ (with O)

$\downarrow$ H_3O^+ / H_2O

H_2O

$\overset{Me_{\delta+}}{\underset{Me}{}}C{-}CH_2$ $\overset{O^{\delta+}}{\underset{H}{}}$

$\downarrow$

$\overset{HO}{\underset{Me}{}}\overset{Me}{\overset{}{}}\overset{H_2}{C{-}C}\overset{}{\underset{OH}{}}$ rotation

$\downarrow$

$Me_2\underset{HO}{\overset{OH}{C}}{-}CH_2$

11 Amines

Aromatic amines are covered in another Primer (M Sainsbury, 'Aromatic Chemistry'). They differ in some important respects from the aliphatic members, the subject of the present chapter.

General tendencies are in scheme **11Ge**.

11Ge

[many formulae for amines etc in the schemes have 2 or more R groups; these may be identical or different]

acts as nucleophile (strong) and base (medium)

acts as very weak acid pK_a ~ 37

N—H

very susceptible to oxidation

Structural types

R—N (H, H)	R—N (R, H)	R—N (R, R)	R—N^+—R Hal$^-$ (R, R)
pm	se	te	quaternary ammonium salts

2.1Re1g; 5Pr2c

Inversion

sp^3 hybridisation

unshared electron pair

σ bonds

106°

enantiomers

at 25° k is between
10^4 s^{-1} (te amines) and
10^{10} s^{-1} (ammonia)

Basicity

e.g. in gas phase

pK_a (of conjugate acid) in H$_2$O

Me$_3$N	>	Me$_2$NH	>	MeNH$_2$	>	NH$_3$
9.8		10.8		10.6		9.2

most important bonding

$-\overset{|}{\underset{|}{N}}\!\!\overset{+}{}-H$ ----- OH$_2$

$H_3C-\overset{4}{C}-\overset{3}{\underset{H_2}{C}}-\overset{2}{\underset{|}{C}}-NH_2$ with $^1CH_3$ and CH_3

2–methylbutan–2–amine

although is an α te R group the compound is a pm amine

$Cl-\underset{H_2}{C}-\underset{H_2}{C}-\underset{H}{N}-CH_3$

2–chloro–*N*–methylethanamine

$HO-\underset{H_2}{C}-\underset{H_2}{C}-NH_2$

2–aminoethanol

$H_2N-(CH_2)_6-NH_2$

hexane–1,6–diamine

With the previous functional groups (e.g. Hal, OH) the terms primary, secondary and tertiary refer to the degree of substitution of the αC. The amino group is different. Here primary, secondary and tertiary signify the number of alkyl groups on the functional group itself (scheme). When the nitrogen is attached to three different atoms or groups the amine is, at any particular instant, a chiral molecule. However amines do not exhibit optical activity. The rapid inversion about the nitrogen interconverts enantiomers and is in effect a fast racemisation. In the gas phase the basicity of amines follows the order expected from the –I effect of the R groups (scheme). In water there is a second important feature, stabilisation of the cations by the hydrogen bonding shown. This works in the opposite sense: NH$_4^+$ should be the most stable cation and hence NH$_3$ the strongest base. The outcome is that, in water, ammonia and the three methylamines do not differ much in basicity. *Dimethylamine* (*N*–methylmethanamine) happens to be the strongest base.

11Pr

1 R–CO$_2$H and derivatives, rearrangement with loss of C

$$R–CO_2H \xrightarrow{\text{8.2Re7}} \quad R–NH_2 \xleftarrow{\text{8.5Re3}} R–CO–NH_2$$
$$R–CO_2R^1 \xrightarrow[\text{8.4Re2b}]{} \text{pm}$$

2 Reduction of functional groups containing nitrogen

a R–CO–NH$_2$ $\xrightarrow[\text{8.1Ge2}]{\text{LiAlH}_4}$ R–CH$_2$–NH$_2$ $\xrightarrow[\text{8.3Re1a}]{\text{R}^1\text{–CO–Cl}}$ R–CH$_2$–NH–CO–R^1

$$\downarrow \text{LiAlH}_4$$

R–CH$_2$–N$\begin{smallmatrix}CH_2–R^2\\[2pt]\ \ \ \ CH_2–R^1\end{smallmatrix}$ $\xleftarrow{\text{LiAlH}_4}$ R–CH$_2$–N$\begin{smallmatrix}CO–R^2\\[2pt]\ \ \ CH_2–R^1\end{smallmatrix}$ $\xleftarrow{\text{R}^2\text{–CO–Cl}}$ R–CH$_2$–NH–CH$_2$–R^1

te   se

b R–N$_3$ ⟶ R–NH$_2$ **c R–NO$_2$** ⟶ R–NH$_2$
azides, **2.1Re1i**   nitro compounds

d R–CN ⟶ R–CH$_2$–NH$_2$ **e** C=N ⟶ CH–NH$_2$
nitriles, **2.1Re1k**   OH
oximes, **7.4Re2c(i)**

LiAlH$_4$ or Pt (or Ni) / H$_2$

3 R–Hal
a with NH$_3$ 2.1Re1f   may react with R–Br

R–Br $\xrightarrow{\text{NH}_3}$ R–NH$_2$ ⟶ R$_2$NH ⟶ R$_3$N ⟶ R$_4$N$^+$ Br$^-$
(best pm)

NH$_3$ + R–Br (excess) $\xrightarrow{\text{high yield}}$ R$_4$N$^+$ Br$^-$ $\xleftarrow{\text{high yield}}$ R$_3$N + R–Br

but NH$_3$ (excess) + R–Br ⤑ R–NH$_2$ yield variable, mixture of products

b with a protected amino group   Gabriel

phthalimide $\xrightarrow{\text{KOH / EtOH}}$ N$^-$ K$^+$
pK$_a$ 9.2

phthalimide (cheap; 1 kg, £10)
benzene-1,2-dicarboximide

R–Br / DMF / Δ
(best pm)

R–NH$_2$ + [diamide] $\xleftarrow{\text{H}_2\text{N–NH}_2\text{ / }\Delta}$ N–R
basic ⤑   neutral

separate products using dilute HCl

4 Amines, reductive alkylation

overall $\overset{\backslash}{\underset{/}{C}}=O$ + $H-\overset{/}{\underset{\backslash}{N}}$ $\xrightarrow[\text{H–CO}_2\text{H or Na}^+\text{H}_3\bar{\text{B}}-\text{CN}]{\text{reduction with Ni / H}_2\text{ or}}$ $\overset{|}{\underset{H}{\overset{\backslash}{\underset{/}{C}}}}\overset{\overset{|}{N}\backslash}{}$

e.g.s

a

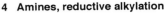

as in **7.4Re2c(lll)**   iminium ion, strong E⁺

$$R_2^1C=O$$
$$NH_3 \xrightarrow{} R_2^1CH-NH_2$$
$$\text{pm}$$
$$RNH_2 \xrightarrow{} R_2^1CH-NHR$$
$$\text{pm} \qquad \text{se}$$
$$R_2NH \xrightarrow{} R_2^1CH-NR_2$$
$$\text{se} \qquad \text{te}$$

b

Me–NH₂ / H–CO₂H / Δ   Leuckart   hydrolysis

as in a   H–CO₂H

#5 tertiary Alcohols or Alkenes   Ritter

e.g. Buᵗ–OH $\xrightarrow[\text{2 H}_2\text{O}]{\text{1 RCN / H}_2\text{SO}_4}$ Buᵗ–NH–CO–R $\xrightarrow{\text{hydrolysis}}$ Buᵗ–NH₂
te pm

Buᵗ–OH₂⁺ ⟶ H₂O + (Buᵗ)⁺ N≡C–R ⟶ Buᵗ–N≡C–R ⟶ $\underset{\text{OH}}{\overset{\text{Bu}^t}{\underset{\backslash}{N}}}=C\overset{R}{\underset{\backslash}{}}$

or Me₂C=CH₂

Of the preparations in scheme **11.Pr** the first, **Pr1**, gives primary amines in high yield. Reduction of amides(**Pr2a**) can be manipulated as shown to produce primary, secondary and tertiary amines. Reduction of other groups (**2b,c,d,e**) is restricted to primary amines.

Reaction of R–Hal with NH_3(**3a**) is bedevilled by the possibility of further reactions (scheme). It does not provide a good general route to primary amines. Protection of the amine group(**3b**) circumvents this difficulty: the alkyl halide gives a substituted phthalimide which does not react further. Hydrolysis of the phthalimide by acid or base is difficult. Treatment with hydrazine liberates the amine which is easily separated from the other product.

Reductive alkylation(**Re4**) in various guises leads efficiently to a range of amines. An excess of methanal with methanoic acid as reducing agent (Eschweiler reaction, **4a**) replaces N–H by N–Me. Methylation of a primary amine cannot be stopped at the mono N–methyl stage. In the Leuckart reaction(**4b**) one N–H of a primary amine is replaced by the C skeleton of the ketone, and the secondary amine is trapped as the amide. (The reaction can be used for converting a secondary to a tertiary amine.) Thus N–methylcyclopentanamine is readily prepared by **4b** but not by **4a**. The common feature is the formation of iminium ions which are reduced by the methanoate anion. [#]Reaction **5** gives primary amines having tertiary alkyl groups. These could not be prepared by, for example, **3b:** a tertiary alkyl halide would undergo elimination.

Aromatic systems are easily nitrated, and the products are usefully converted to other functional groups directly attached to aromatic rings. Aliphatic nitro compounds, which are less readily accessible, are not so important. Nevertheless the simplest member, nitromethane, is a cheap commercial product (2kg, £36) and is involved in several synthetic sequences. Its reactions are summarised at the start of scheme **11Re**, the rest of which deals with the reactions of amines.

11Re

2 With R¹–CO–CI   8.3Re1a,b

3 With HNO_2

a primary amines

HCl / $NaNO_2$ / H_2O / 5°

usually represented as $R–NH_2$ + HNO_2 = $R–OH$ + N_2 + H_2O

but following example illustrates complexity of reaction

$H–O–N=O$ $\xrightarrow{H^+}$ $H_2\overset{+}{O}–N=O$ $\longrightarrow$ $\overset{+}{N}=O$ + Pr$–\ddot{N}H_2$

nitrous acid   nitrosonium ion, E^+ and loss of H^+

Pr$–\overset{+}{N}{\equiv}N$ $\longleftarrow$ Pr$–\overset{+}{N}{\equiv}N\overset{\frown}{–}\overset{+}{O}H_2$ $\longleftarrow$ Pr$–N=N–O–H$ $\longleftarrow$ Pr$–\underset{H}{N}–N=O$

diazonium ion

$H_3C–\underset{H_2}{C}–\overset{+}{C}H_2$ or $H_3C–\underset{\underset{H}{|}}{\overset{\overset{H}{|}}{C}}–\overset{+}{C}H_2$ or $H_3C–\underset{\underset{H}{|}}{\overset{\overset{H}{|}}{C}}\overset{+}{–}CH_2$ or   (cyclopropane cation with H_2C, $\overset{+}{C}H_2$, $\underset{H_2}{C}$)

│H_2O │ migration of H^- │ loss of H^+ │ loss of H^+

$H_3C–\underset{H}{\overset{+}{C}}–CH_3$

│H_2O loss of H^+

$H_3C–\underset{H_2}{C}–\underset{H_2}{C}–OH$ + $H_3C–\underset{\underset{OH}{|}}{\overset{\overset{H}{|}}{C}}–CH_3$ + $H_3C–\underset{H}{C}=CH_2$ + (cyclopropane)

b secondary amines

$R_2N–H$ $\xrightarrow{HNO_2}$ $R_2N–N=O$ nitrosamines; yellow oils, so weakly basic that are insoluble in dilute HCl

c tertiary amines

R_3N react very slowly with HNO_2 to give a complex mixture of products

4 Oxidation

5Re3c

R_3N $\xrightarrow{Ar–CO_3H \,/\, HCCl_3}$ $R_3\overset{+}{N}–\overset{-}{O}$ + $Ar–CO_2H$

amine oxides

$\underset{HO}{\overset{\delta+}{}}\overset{\delta-}{}\underset{}{\overset{O}{\overset{||}{O–C}}}–Ar$ $\longrightarrow$ $R_3\overset{+}{N}–\overset{-}{O}\overset{\frown}{H}$ + $\overset{-}{O}–\overset{O}{\overset{||}{C}}–Ar$

5 With dichlorocarbene

$R–NH_2$ $\xrightarrow{HCCl_3 \,/\, NaOH \,/\, H_2O \,/\, \Delta}$ $R–\overset{+}{N}{\equiv}\overset{-}{C}$

isonitriles **2.1Re1k**

fast

$\overset{\frown}{Cl}–\overset{-}{C}Cl_2$

slow

loss of 2 HCl

(resonance structures of dichlorocarbene) $H_2\ddot{N}–R$ $\longrightarrow$ $Cl_2\overset{-}{C}–\underset{H_2}{\overset{+}{N}}–R$

dichlorocarbene

#6 With methanal and an enolisable ketone

Mannich

overall e.g. $Me_2\overset{+}{N}H_2$ $\overset{-}{Cl}$ + CH_2O + $Ph–CO–CH_3$ $\longrightarrow$ $Ph–CO–CH_2–CH_2–NM$

$Me_2N–H$ $\xrightarrow{\text{as in } \textbf{11Pr4}}$ $Me_2\overset{+}{N}=CH_2$ $\quad H_2C=C\overset{O-H}{\underset{Ph}{}}$

$\overset{+}{H}\overset{-}{Cl}$

HCl

at pH 7.5

tropinone

+ 2 CO_2 + 2 H_2

cyclohexanone

+ $H_3C–NO_2$

1 NaOEt / EtOH
2 AcOH

$HO\ \ \overset{H_2}{C}\!-\!NO_2$

H_2 / Ni / AcOH

$HO\ \ \overset{H_2}{C}\!-\!NH_2$

filter, add ice,
then NaNO_2

$H{-}O\ \ \overset{H_2}{C}\!-\!N{\equiv}N$

Tiffeneau rearrangement

cycloheptanone
(overall yield 42%)

Nitromethane gives a stable anion(**Re1**) which serves as a component of aldol condensations. Acidification leads to the less stable aci tautomer. The remarkable feature is that reversion to the nitro form is *slow*. Thus by careful work it is possible to generate the aci form in high concentration; a few aci tautomers have been isolated.

The reactions of nitrous acid with amines(**Re2**) are notoriously dirty; simple amines give a plethora of products (scheme). Although the nitrosonium ion is shown as the electrophile other species (e.g. N_2O_3) may also be involved. The sequence of reactions leads to unstable carbocations. These are so reactive that there is little selectivity in their reactions. This accords with the general principle(**4.2Re1c**). Paradoxically, with certain more complex amines the reaction is clean and preparatively useful. An example is given later.

Oxidation of amines, an extremely complicated subject, is famous in the annals of organic chemistry. In 1856 W H Perkin (then aged 18) oxidised impure aniline (phenylamine) and obtained mauveine, the first synthetic dye. This spawned a huge chemical dyestuff industry. Only one reaction is included here, the preparation of amine oxides from tertiary amines(**Re4**).

Although the yields in **Re5** are usually modest this route to isonitriles is better than the alternative of treating alkyl halides with silver cyanide.

#The Mannich reaction(**Re6**) generally involves a secondary amine, methanal or a few other reactive aldehydes, and a ketone which can enolise. The yield increases in line with the enol content of the ketone, as would be expected from the mechanism shown. A striking example is the synthesis of tropinone, an alkaloid derivative, by Robinson in 1917.

Nitromethane is the key component in the ring expansion of cyclohexanone (margin). The nitrous acid reaction probably does not go so far as a carbocation. Loss of nitrogen from the diazonium ion is accompanied by migration of a C–C bond; this migration is assisted by loss of a proton and formation of a carbonyl group. At first sight the overall yield does not seem impressive. However there are in effect four or five stages, and each of them must proceed in a yield higher than 80%.